'Petroleum economies have long been analyzed macro-economically as suffering from the "resource curse". This book offers a long-awaited alternative view based on a knowledge economy perspective. Petroleum economies can benefit from complex knowledge built up in supplier industries to diversify into new and promising industries. The case of Norway, central to this book, serves as an example for many other resource-based economies worldwide.'

Koen Frenken, *Professor in Innovation Studies*, *Utrecht University*, *Netherlands*

'This book reveals the dynamics of natural resources when developed by a capability rich institutional regime. The petroleum sector in Norway is not only a success story, it has also transformed the innovation models of the global petroleum industry. The book is a must for those who want to understand today´s offshore industry as well as for those who want to prepare for the transitions to come.'

Staffan Laestadius, *Professor Emeritus*, *Royal Institute of Technology*, *Stockholm*, *Sweden*

'The oil and gas industry remains the largest in the world by a long way: it is the resource at the heart of the industrial system. This path-breaking book explores one of its most dynamic national bases [Norway]. It offers unique insights into oil's innovation paths, its industrial trajectories, its economic impacts and its future.'

Keith Smith, *Professor at Imperial Business School*, *UK*

Petroleum Industry Transformations

Taking the case of the Norwegian petroleum industry as its vantage point, the book discusses the question of industrial transformations in resource-based industries. The book presents new, empirically-based analyses of the development of the petroleum industry, with an emphasis on three ongoing transformation processes:

- Technological upgrading and innovation in upstream petroleum.
- Globalisation of the petroleum industry and suppliers' experiences of entering foreign markets.
- Diversification into and out of petroleum – and the potential for new growth paths after oil.

Drawing together a range of key thinkers in this field, this volume addresses the ways in which the petroleum industry and its supply industry has changed since the turn of the millennium. It provides recommendations for the development of resource economies in general and petroleum economies in particular.

This book will be of great interest to students and scholars of energy policy and economics, natural resource management, innovation studies and the politics of the oil and gas sector.

Taran Thune is Professor in the Center for Technology, Innovation and Culture at the University of Oslo, Norway.

Ole Andreas Engen is a Professor at the University of Stavanger, Norway.

Olav Wicken is Professor in the Center for Technology, Innovation and Culture at the University of Oslo, Norway.

Routledge Studies in Energy Transitions

Series Editor: Dr. Kathleen Araújo, *Stony Brook University*, USA

Considerable interest exists today in energy transitions. Whether one looks at diverse efforts to decarbonize, or strategies to improve the access levels, security and innovation in energy systems, one finds that change in energy systems is a prime priority.

Routledge Studies in Energy Transitions aims to advance the thinking which underlies these efforts. The series connects distinct lines of inquiry from planning and policy, engineering and the natural sciences, history of technology, STS, and management. In doing so, it provides primary references that function like a set of international, technical meetings. Single and co-authored monographs are welcome, as well as edited volumes relating to themes, like resilience and system risk.

Titles in this series include:

Petroleum Industry Transformations
Lessons from Norway and Beyond
Edited by Taran Thune, Ole Andreas Engen and Olav Wicken

Petroleum Industry Transformations

Lessons from Norway and Beyond

Edited by Taran Thune, Ole Andreas Engen and Olav Wicken

Routledge
Taylor & Francis Group
LONDON AND NEW YORK

earthscan
from Routledge

First published 2019 by Routledge

2 Park Square, Milton Park, Abingdon, Oxfordshire OX14 4RN
52 Vanderbilt Avenue, New York, NY 10017

Routledge is an imprint of the Taylor & Francis Group, an informa business

First issued in paperback 2020

British Library Cataloguing-in-Publication Data
A catalogue record for this book is available from the British Library

Library of Congress Cataloging-in-Publication Data
Names: Thune, Taran, editor. | Engen, Ole Andreas, editor. | Wicken, Olav, editor.
Title: Petroleum industry transformations: lessons from Norway and beyond/edited by Taran Thune, Ole Andreas Engen, Olav Wicken.
Description: Abingdon, Oxon; NewYork, NY: Routledge, 2019. |
Series: Routledge studies in energy transitions
Identifiers: LCCN 2018014688 | ISBN 9781138307636 (hardback) |
ISBN 9781315142456 (ebook)
Subjects: LCSH: Petroleum industry and trade–Norway. | Economic development–Norway.
Classification: LCC HD9575.N62 P48 2019 | DDC 338.2/72809481–dc23
LC record available at https://lccn.loc.gov/2018014688

ISBN: 978-1-138-30763-6 (hbk)
ISBN: 978-0-367-51879-0 (pbk)

Typeset in Goudy
by Wearset Ltd, Boldon, Tyne and Wear

Contents

PART IV

Figures

Tables

Contributors

Allan Dahl Andersen, Researcher, Centre for technology, innovation and culture. University of Oslo

Atle Blomgren, Senior Research Scientist, Norwegian Research Centre AS (Norce)

Antonio José Junqueira Botelho, Professor, Graduate Program in Political Sociology, University Cândido Mendes, Brazil

Ole Andreas Engen, Professor, University of Stavanger

Rune Dahl Fitjar, Professor, Centre for Innovation Research, University of Stavanger

Jakoba Sraml Gonzalez, PhD Fellow, Centre for technology, innovation and culture. University of Oslo

Magnus Gulbrandsen, Professor, Centre for technology, innovation and culture. University of Oslo

Gary Herrigel, Paul Klapper Professor, Department of Political Science, University of Chicago

Nina Hjertvikrem, PhD Fellow, Centre for Innovation Research University of Stavanger

Tuukka Mäkitie, PhD Fellow, Centre for technology, innovation and culture, University of Oslo

Petter Osmundsen, Professor, University of Stavanger

Øystein Noreng, Professor Emeritus, Norwegian School of Management BI

Christian Quale, Senior Advisor, Norwegian Research Centre AS (Norce)

Helge Ryggvik, Professor, Centre for technology, innovation and culture, University of Oslo

Charles Sabel, Maurice T. Moore Professor of Law, Columbia Law School, Columbia University

Erlend Osland Simensen, PhD Fellow, Centre for technology, innovation and culture. University of Oslo

Taran Thune, Professor, Centre for technology, innovation and culture, University of Oslo

Olav Wicken, Professor, Centre for technology, innovation and culture, University of Oslo

Acknowledgements

This book developed out of a research project called SIVAC (Supplier industry and value creation) supported by the Norwegian Research Council and the Petrosam2/Petromaks programs (project no. 237677). Their economic support and interest in this research project is gratefully acknowledged.

Several of the book chapters have been presented and discussed at meetings, seminars and conferences in Norway and elsewhere. Insightful comments and constructive criticism from our colleagues in the SIVAC project, in our departments and in the Norwegian and international research community in innovation management, resource economics and industrial transformations is acknowledged as well.

All authors and contributors to this book thank a vast number of specialists in petroleum-related government agencies, the oil industry and the supply sector for providing information and access to data, as well as suggestions and comments on chapters and presentations.

In the final stages of completing the book, we received support from John Taylor and Robin Fiske, as well as from the editorial staff in Routledge, particularly Matt Shobbrook. We thank you for helping us in keeping the book on track.

Finally, the editors thank all the contributors to the book, who in addition to providing interesting and insightful chapters, did not complain too much about our short deadlines, excessive demands and constant nagging. This book would never have materialized without you!

Taran Thune, Ole Andreas Engen and Olav Wicken (the editors)

Abbreviations

APA	Allocation in Predefined Areas
CCS	Carbon, Capture and Storage
CEO	Chief Executive Officer
CIS	Community Innovation Survey
CoPS	Complex Product Systems
CTO	Chief Technology Officer
DUI	Doing, Using and Interacting
E&P	Exploration and Production
EEA	European Economic Area
EPC	Engineering, Procurement and Construction
EPCI	Engineering, Procurement, Construction and Installation
FEED	Front-End Engineering and Design
FMC	Food, Machinery and Chemicals
FPSO	Floating Production, Storage and Offloading
FPUs	Floating (semi-submersible) Production Units
GoM	Gulf of Mexico
IOC	International Oil Company
NACE	Nomenclature statistique des Activités économiques dans la Communauté Européenne
NCS	Norwegian Continental Shelf
NHO	Confederation of Norwegian Industry
NOC	National Oil Company
NOK	Norwegian Kroner
O&G	Oil and Gas
OPEC	Organisation of Petroleum Exporting Countries
OWP	Offshore Wind Power
PL	Production Licence
PLET	Pipeline End Termination
PLSV	Pipelay Support Vessel
R&D	Research and Development
ROVs	Remotely Operated Vehicles
SDFI	State Direct Financial Involvement
SIVAC	Supplier Industry and Value Creation

SSB	Statistics Norway
SSTB	Subsea Tie-Back
STI	Science, Technology and Innovation
SWRI	Supplier Working Relation Index
TLP	Tension-Leg Platform
US$	US dollars

1 Transformations in petroleum

Innovation, globalisation and diversification

Taran Thune, Ole Andreas Engen and Olav Wicken

Natural resource industries – innovative and dynamic?

This book addresses the question – to what extent are resource industries dynamic and innovative, particularly in the context of the upstream petroleum sector. A significant literature argues that economies with an abundance of natural resources – such as petroleum economies – are characterised by a 'resource curse' (Sachs and Warner, 1995, 2001). This generally means that a boom in a natural resource industry causes high financial flows into the economy, resulting in competition among both political and economic elites to access what are known as 'resource rents'. Resource rents, or the value of capital flows rendered by exploiting natural resources, are particularly high in the exploitation of valuable natural resources such as oil and gas. In turn, the effect over time is that labour and resources will move away from manufacturing and other industries that compete in open markets and move mainly towards services that are less exposed to international competition. Implicit in the resource curse concept or model is the idea that the inherent processes will move resources away from the more innovative parts of the economy, such as manufacturing, towards less dynamic areas, namely natural resources and services. Following this logic, natural resource-based economies are assumed to experience lower growth and rapid deindustrialisation (Corden and Neary, 1982; Sachs and Warner, 1995, 2001).

There has been considerable discussion of the resource curse hypothesis. An emerging strand in the literature on industrial development argues that resource economies are not necessarily characterised by the 'curse'. Focusing on knowledge instead of financial flows, the alternative view highlights innovation processes and industrial dynamics within the natural resource industries (David and Wright, 1997; Ville and Wicken, 2012; Andersen *et al.*, 2015; Mahroum and Al-Saleh, 2016; Wicken, 2016). This alternative view does not assume a priori that all natural resource industries are 'low tech' or have low levels of innovation (von Tunzelman and Acha, 2005). Rather, such writing emphasises differences between natural resource industries and across different countries in terms of their ability to improve productivity, increase production volume and develop new products and markets. Moreover, this newer perspective argues that in order to understand the dynamism of resource industries, one also has to look at the linkages between resource-based sectors and the rest of the economy (Andersen, 2012; Morris, Kaplinsky and Kaplan, 2012). In such work, the emphasis has been on the interaction between technologically advanced 'enabling sectors' and the 'recipient' resource sectors (Ville and Wicken, 2012; Andersen *et al.*, 2015). Resource sectors, however, cannot be passive recipients; they need high levels of absorptive capacity since selection and integration of advanced solutions occurs within large and complex technological systems.

With the above perspective as a backdrop, this book addresses an important natural resource sector – the upstream petroleum industry. In the case of the petroleum industry, the enabling sector would be those companies supplying the oil and gas companies with equipment, technology and services. The ambition is to explore how the relationship between the enabling supply industry and the recipient petroleum sector has given rise to an innovative and dynamic natural resource industry. The empirical case studied here is the Norwegian petroleum industry which we regard as a good case for addressing the innovative capability and development of the petroleum sector.

The upstream petroleum sector: actors, collaboration and specialisation

To understand the development of the upstream petroleum sector over the last decades, it is necessary to understand that the industry works as a complex industrial–political system with multiple actors, technologies and institutional contexts. Broadly defined, three sets of actors are involved in this system: upstream petroleum companies (occasionally referred to as oil companies or operators), petroleum supply companies (also sometimes referred to as oil service companies) and public sector organisations which regulate and support the industry. The two major kinds of companies in the upstream industry (the operators and service and supply companies) are largely complementary, but have different types of assets and business models. The petroleum supply industry can be defined as companies that 'provide assets, equipment, technology, manpower and project management that enable oil companies … to explore and develop oil and gas fields' (Beyazay-Odemis, 2016, p. 25). Upstream petroleum companies, on the other hand, are companies which own and operate licences for searching, drilling, producing and transporting crude oil and gas from underground or undersea deposits to the surface, and onshore for further refinement. Upstream petroleum companies are occasionally referred to as exploration and production companies.

Until a few decades ago, large, integrated petroleum companies such as Exxon, Shell, BP and Chevron developed technologies for the extraction of oil and gas and then deployed them in different petroleum regions across the globe. During the 1980s and 1990s, increased volatility in oil prices, technological risks and complexity of operations, and the necessity to deploy advanced communication technologies, led to increased need for specialisation as well as collaboration between suppliers and oil companies. The increased nationalisation of petroleum resources in different regions was also important. According to several analysts, the interdependency between petroleum companies and their suppliers has become even greater during recent decades. Petroleum companies have increasingly disintegrated their operations, and have moved activities out of oil companies and have created a complex web of suppliers around them (Acha, 2002; Shuen *et al.*, 2014; Bagheri and Di Minin, 2015). Further, new, national oil companies and smaller independent oil companies have fewer resources to develop technologies. To understand the innovative capability and prospects for the petroleum industry, one needs a better understanding of the dynamic relationship between the main players in the ecosystem (Acha, 2002).

Currently, most oil operators concentrate on identifying and characterising reservoirs and the overall management of the exploration and extraction processes. Moreover, the oil companies function as system integrators and lead users, and have an overall responsibility for deploying technologies in the upstream value chain. Both in the development and operation phases of petroleum fields, a vast array of different technologies are used, and need to function as part of a system. Thus, in addition to assessing the performance and safety of individual solutions, new technologies have to perform alongside a range of others. The petroleum companies and their main suppliers, so-called EPCI or engineering, production, construction and installation companies, therefore require broad technological competence, not only within their own core assets.

The upstream industry comprises two, rather than one, dominant technological regimes (Acha, 2002). Operator companies have substantial knowledge of how to identify and manage hydrocarbon reserves. Most of this knowledge is synthetic in character and derives from operational experiences in diverse geological settings. Knowledge that is of strategic advantage for oil operators is not of the technological applications per se, but rather the knowledge of how and where to deploy them. For suppliers, development and deployment of technologies is their core business. Protecting rights to technologies is central for suppliers, whereas operators are often more willing to share technology they have developed internally, so that other companies can exploit it commercially (Perrons, 2014).

Among the supply firms there is a system hierarchy, and at the top are so-called 'service supermajors' (Acha, 2002) such as Halliburton and Schlumberger. These companies operate as the main contract partners to the oil companies in field development and maintenance work on existing fields. They work with a long list of sub-suppliers which deliver the tools and services needed to develop and manage petroleum fields onshore and offshore. Many supply companies operate in specific segments of the petroleum value chain, from seismic services used for locating hydrocarbon reserves offshore, to firms only involved in plugging wells and decommissioning old oil fields. The technology intensiveness of suppliers also varies considerably according to the part of the value chain in which they are positioned, their size and maturity.

Compared to other industrial sectors that are characterised by distributed value chains and collaboration with suppliers (see Chapter 15), the petroleum companies maintain a high degree of control over their suppliers and collaborate closely with them. As described above, exploring and producing petroleum requires a vast number of technologies that need to function as an integrated system. Moreover, the solutions required are often unique to the geophysical area where the solutions are to be implemented. This requires a high degree of specialised knowledge about the natural conditions as well as broad technological competencies. Petroleum companies are therefore unique users of innovations, and are very much involved in all stages of technology development including the provision of funds for exploratory work. The necessity of integrating multiple solutions from different vendors means that there is strong emphasis on industry-wide standards and qualification of technologies, but individual oil companies and field development projects will also have unique requirements for technologies and solutions that are fitting for specific technological purposes and geological conditions.

The empirical context – the Norwegian petroleum sector

This book extends coverage of the above issues based on several empirical studies of industrial dynamics and innovation in petroleum. It does so with a focus mainly on the North Sea petroleum region since the turn of the millennium.

Oil was first discovered in the North Sea region in the 1960s. The region entered into what is defined as a mature phase characterised by diminishing new large-scale discoveries, stagnation in production and increased costs in extraction of existing oil and gas in the 1990s. The exploration of mature petroleum regions, where most of the 'easy oil' has been extracted, requires increased investment in advanced production technologies for enhanced recovery and production of more complicated fields. This, in turn, requires greater innovation activity. The result was a rapid increase in investment and demand for services on the Norwegian shelf, indicated by the rise from 60 to 226 billion NOK between 2001 and 2014, which was subsequently reduced to 166 billion NOK in 2016 (Statistics Norway, 2018).

The period from the late 1990s until 2014 was characterised by a significant increase in prices of petroleum products, particularly crude oil. The prices reached an all-time high during that timeframe, with a peak of about 145 dollars per barrel in 2008. Neither before nor since in the world history of the petroleum industry have petroleum prices reached such a level (U.S Energy Information Administration, 2018). Following a decline in 2009 to 40 dollars per barrel, prices again climbed to an average of 100 dollars in the years 2010–2015. They then fell again to between 40 and 50 dollars per barrel in 2016–2017. The activity level of the industry is highly sensitive to fluctuations in petroleum prices, where small variations may cause changes in financial decisions, budgets and other cost-related parameters.

Due to increased maturation and consistently high prices for oil, petroleum production reached a historic peak as part of Norway's national economy during recent years. In 2014, the upstream petroleum sector contributed to 22 per cent of GDP, 60 per cent of total exports of goods and more than 50 per cent of total exports as well as 30 per cent of total investments and a third of all state revenues (Figure 1.1).

During the last two decades, the Norwegian economy became strongly dependent on natural resources, particularly petroleum, but also emerged as one of the wealthiest countries in the world. At the same time, the country is seen (and sees itself) as one of the rare cases of a natural-resource-specialised country that has been able to escape the resource curse and has supported the build-up of broad industrial capabilities at the same time as becoming a petroleum nation (Gylfason, 2001, 2004). The manufacturing sector in Norway experienced a significantly higher rate of growth compared to other industrial economies in Northern Europe between 1990 and 2014 (Wicken, 2016), as illustrated in Figure 1.2.

An important aspect of the relatively successful industrial development in Norway is the rapid growth of the petroleum supply sector which to a large extent has compensated for de-industrialisation in other parts of the manufacturing sector.

If the supply industry was defined as a stand-alone, industrial sector, it would be Norway's second largest, following oil and gas. Total sales by the industry increased from 140 billion NOK in 2001 to 357 billion in 2010, and reached a peak of 520 billion NOK in 2013. The downturn in the industry had a significant impact as total sales dropped to 378 billion NOK in 2016 (Rystad Energi,

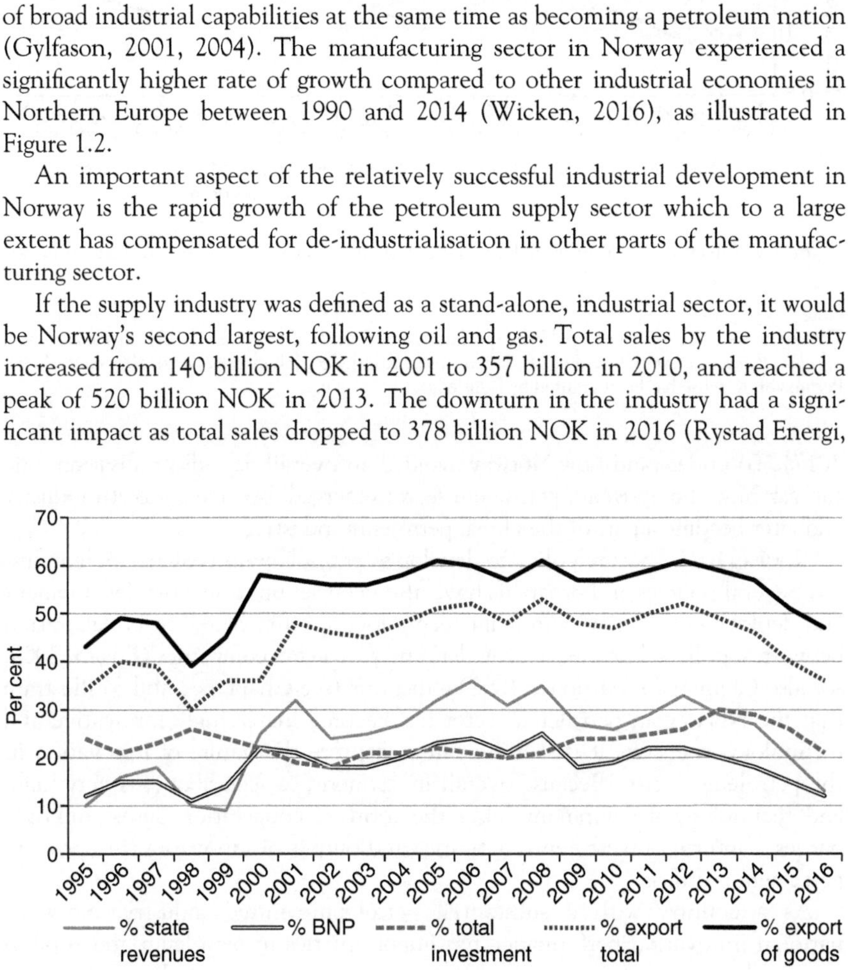

Figure 1.1 Macroeconomic indicators for the petroleum sector, 1971–2017.

Source: Statistics Norway (National accounts), Ministry of Finance (The national budget 2018). Available at www.norskpetroleum.no/en/economy/governments-revenues (retrieved 07.06.18).

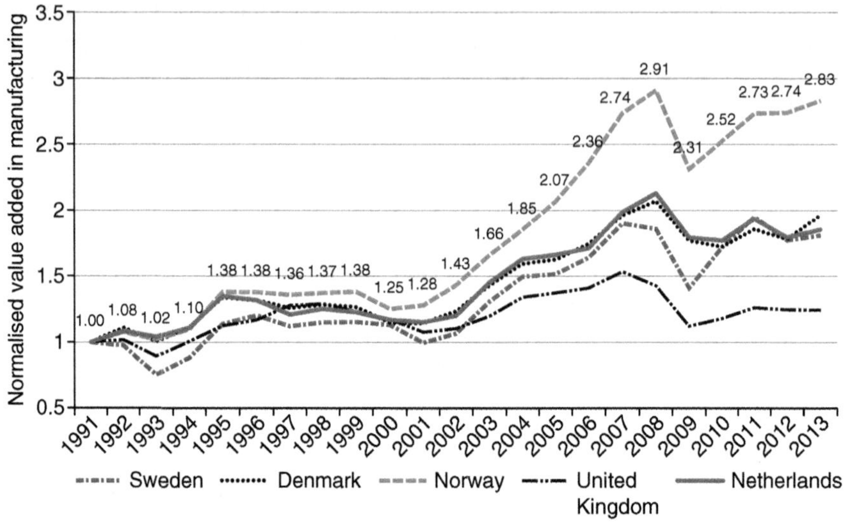

Figure 1.2 Growth in value added in manufacturing sector in selected industrial countries, 1991–2014.

Note
This figure has previously been published in Mahroum, S. and Al-Saleh, Y. (2017): *Economic Diversification Policies in Natural Resource Rich Economies*. Routledge, (chapter 12, page 300, figure 12.3). Permission to reuse has been granted by Routledge.

2017). To understand how Norway avoided an overall de-industrialisation, one can ask how the upstream petroleum sector emerged, became a growth industry and later become a part of the global petroleum industry.

Norway has a specific policy to develop a technology-intensive-supply industry. Several policies and programs have also been set up to support development and deployment of new petroleum technologies, often based on collaboration between suppliers, knowledge providers and petroleum companies (Engen, 2009; see also Chapter 2). Although R&D indicators (see Chapters 2 and 3) illustrate that the Norwegian petroleum sector has become increasingly innovative and technology-intensive, it is still difficult to capture the nature of innovation in the petroleum sector. Because overall investment, as seen above, is very high, and technology development takes the form of cooperation across multiple parties, capturing investments, activities and output of innovation in upstream petroleum is difficult.

As a country with a substantial petroleum-centred industry, Norway's national innovation performance has important ties to petroleum and is often portrayed as a mediocre performer compared to other northern European countries, such as Germany, Denmark or Sweden. By international statistical standards, resource- and commodity-based industries are usually regarded as 'low-' or 'medium-tech', which certainly also includes the oil and gas industry

– an industry that has been found to have an R&D intensity which is less than 1 per cent of its overall expenditure (von Tunzelmann and Acha, 2005). This image contrasts with the common perception of an increasingly technology-intensive and innovative oil and gas sector in Norway, and has led several analysts to conclude that the nature of innovation in the oil and gas sector is not captured by standard technology indicators such as R&D expenditure. More detailed empirical studies of innovation in the oil and gas industry are necessary, and to which a section of this book is devoted.

The goal of the book is to provide a new and empirically-based perspective on transitions in the petroleum industry. Empirically, we look into the transformation of the upstream petroleum industry over the last two decades and the role of technology-intensive supply companies in this process. Emphasis is placed on changes that have occurred after the turn of the millennium, with retrospective consideration to contextualise and explain recent and ongoing change. The book also has a forward-looking orientation, discussing the opportunities and challenges for diversification of older natural resource industries during phases of declining output. This is an important question for fossil-energy-producing countries during the global energy transition process involving de-carbonisation of energy systems.

Ambition and profile the book

The petroleum supply industry is a major global industry. Recognising this, it is strange that the industry has rarely been studied by researchers interested in industrial development and innovation (Acha, 2002). It is also unusual due to the industry's importance in the world economy, and because of how technologically advanced it is. There are several likely reasons for this condition. The petroleum supply industry is a relatively young and also very complex industry. Due to this, the petroleum sector in general and its supply industry in particular are hard to classify in statistical terms. Petroleum supply companies are not united in the products they produce nor in the services they offer, but rather in the markets that they serve. This means that they are difficult to find in statistical databases of industries and firms. Moreover, existing research on the oil and gas industry has tended not to focus on innovation or technology development and deployment, but rather on the effects of national policy or energy prices on activity levels (Pinder, 2001).

For countries, regions and firms with considerable investment in petroleum, further knowledge of industrial development connected to petroleum is important. In the last five years, the petroleum industry has come under strong pressure due to high costs, lower investment levels, lower commodity prices, lower demand, increasing competition from unconventional hydrocarbon sources and renewables, and changing regulatory and institutional frameworks. Long-term challenges from renewable energy and changed transportation patterns could eventually lead to the phasing out of the entire upstream petroleum industry, but this is not likely to occur in the foreseeable future. However, some petroleum-producing countries, like Norway and the UK, seem to be in the process of bracing themselves for a substantial industrial downturn. In this process, pertinent questions are how dependent these countries are on their petroleum resources and whether it is possible to transfer petroleum-related companies, and particularly suppliers, to non-petroleum-related markets.

To address these issues and to discuss implications for policy and practice, there is a need for updated and systematic knowledge of the petroleum industry, broadly defined. This book is a step in this direction and provides a predominantly empirical answer. It is not based on a singular theoretical perspective or a methodological approach. The different chapters of the book use different data, analyses and theoretical concepts to shed light on the Norwegian petroleum industry case, and combine traditional economic perspectives and analyses of the petroleum sector with innovation theory, network theory, management perspectives, theory on technological styles as well as transition theory. Some of the chapters are also distinctly a-theoretical, providing historical and empirical accounts of the recent development of the Norwegian petroleum industry.

What the chapters have in common, however, is that they emphasise co-evolution between the petroleum producers and the suppliers (as discussed in Part I), and the arrangements and regulations shaping the character of the interaction between the two main groups of industrial actors. These regulations constitute the central institutions for interactive learning and innovation within the sector-specific innovation system. The empirical focus in many chapters is on various specific arrangements developed to influence interaction between firms in different industrial sectors involved in innovation processes, or what can be defined as 'social technologies' (Nelson and Sampat, 2003). Various chapters describe the introduction of specific tools regulating these relationships since the 1990s until today.

The empirical content of the book is predominantly based on data from Norwegian petroleum and supply companies (either in Norway or abroad), but an international perspective is also provided in several chapters in order to foster a discussion of what can be learnt about industrial development in natural resource economies (Chapters 14 and 15 in particular). In this book, the Norwegian development of the petroleum sector is seen in the light of broader development trends in global upstream oil and gas. The key trends can be described as increasingly technology-intensive, increasingly global and increasingly diverse in competencies and applications (Acha, 2002; Inkpen and Moffett, 2011; Perrons, 2014; Bahgeri and Di Minin, 2015).

To reflect on these trends, the book is divided into three main sections as well as a concluding section focusing on international perspectives and lessons learnt. Below, we briefly address the main questions addressed within each section, present the chapters and highlight some key findings on the three main issues.

The four parts of the book and presentation of chapters and main findings

Part I: innovation in upstream oil and gas and the role of suppliers

The first part of the book discusses the industrial development of the petroleum sector by looking at innovation activities. The key perspective in this part of the book is that petroleum constitutes a highly innovative resource-based industry where innovation activities are distributed across multiple players. We position this work in contrast to a traditional perspective on resource economies that regard such industries as low-tech and not very innovative. The key ambition of this part of the book is therefore to describe and discuss the unique innovation model that exists in the upstream sector and how it has developed over time.

As discussed above, innovation and industrial dynamics in upstream petroleum have received less attention from researchers than what they might deserve. One reason is that the petroleum industry is often seen as a slow industry, preoccupied with process improvements and incremental developments (Acha, 2002). As in other engineering-based industries with a synthetic knowledge base, measuring innovation performance and technology intensity is difficult as both data on the investment side and economic performance are of such character that investment and deployment of new technologies are likely to be underreported (von Tunzelman and Acha, 2005).

On the other hand, the general impression of industry experts is that petroleum identification and extraction has increasingly become a high technology 'play', but which perhaps is not reflected in the statistical data. This paradox is part of the motivation behind this book. How can it be that an industry that is highly innovative in the eyes of industry experts is seen by innovation researchers as low-tech and not very advanced? Attempting to understand the innovation model that lies behind this has been important for us.

Among other things, previous research has shown that investment in R&D for development of new technologies is increasingly carried out by suppliers, and testing and implementation of new and exploratory technologies might not be accounted as R&D by the operators. According to Perrons (2014) operators increasingly rely on suppliers to innovate, and mainly implement new innovations developed by suppliers, and supply firms are categorised in other sectors in industrial statistics. Both Perrons (2014) and Acha (2002) claim that over time large service companies have taken over much of the technological frontier as, for instance, measured by the number of patents in upstream technological segments. This observation points to the changing roles of key participants within the sectoral innovation system of the upstream petroleum industry where supply companies increasingly play the role of innovation generators, and operator companies increasingly take on the role of lead user, sponsor and test site for new technologies.

Partnerships between oil operators and with supply companies are important, organised as joint industry projects, commissioned work and collaborative R&D

projects. Supply companies that develop new technologies and offer services on deployed technologies have strong incentives to collaborate with operator companies. In the oil industry, use of demonstration projects and field-testing is necessary for successful commercialisation of new solutions, which means that suppliers are dependent on the operators to get their technologies into the market. The actual design and development of new solutions is executed by suppliers, but the users need matching capabilities to be able to determine their functionality within the overall system. The petroleum companies operationalise demand and set the direction for the innovative efforts carried out in the technology-intensive supply industry. Supply companies compete on developing technologically advanced and cost-competitive products and services for all phases of the petroleum extraction and production processes. In addition to the major players, both operators and service companies require products and services offered by a range of specialised sub-suppliers and public research communities.

With this as a backdrop, Part I of the book presents chapters that all address the issue of innovation in upstream oil and gas, and the role of the supply industry within an increasingly technology-intensive upstream industry. These chapters attempt to understand the nature of innovation, particularly the intricate collaboration between users and producers of technologies, and how they are part of the innovation story of upstream oil and gas. Since innovation activities are the result of collective efforts, the evolving conditions for collaboration is also a key topic, particularly how these are associated with changes in the oil and gas market.

Chapter 2 commences this part of the book, with a broad account of the development and transformation of the sectoral innovation system for upstream oil and gas in Norway. The chapter draws on a sectoral innovation systems framework and discusses developments occurring over the last two decades, focusing on changes in actors, technologies and institutional framework conditions. This chapter looks at the co-evolution between system elements to account for industrial development and changing demand conditions.

Chapter 3 looks further into the issue of modes of innovation in upstream oil and gas, and whether the available statistical data used to compare the R&D and innovation performance across industries adequately represents the innovativeness of the upstream oil and gas industry. Second, this chapter presents a micro-level account of innovation among petroleum and supply companies, and highlights the collaborative nature of innovation. This is further illustrated by a network analysis of research and innovation partnerships. The main finding in this chapter is that innovation performance in oil and gas is mainly a result of collaboration between operators and suppliers, and that such activities are not adequately captured by existing indicators of innovation.

Chapter 4 also addresses the collaborative model of innovation that characterises the upstream industry, and presents a case study of a specific innovation network – a subsea technology network in Rogaland. This study finds a high degree of interaction between members of the network, particularly between the petroleum companies and the supply companies, and also an overall positive effect of networking upon innovation performance.

The study in Chapter 4 characterises innovation activities carried out in periods with very favourable economic conditions with much investment in new technologies. However, the upstream sector – as with all resource-based industries – is characterised by high volatility in commodity prices. It is therefore important to take into account how business cycles dominating the industry influence innovation in upstream oil and gas. We turn to this in the following two chapters.

Chapter 5 analyses the question of suppliers' innovation activities during industry downturns, and whether they are able to sustain innovation activities when users (i.e. petroleum companies) have to radically cut costs. The recent downturn in the industry (2014–2017) is the empirical setting for this comparative case study of multiple suppliers in different parts of the petroleum value chain.

Chapter 6 looks at a particular niche within the petroleum supply industry – the rig service industry, and discusses how innovation and transformation in contractual relations has occurred following the need for oil companies to curb costs. While some of these changes were partly determined by economic conditions, other innovations represented more permanent adjustments to collaborative relations between oil companies and contractors. New oil companies are less keen than the established players to build up a large internal staff to supervise drilling operations. This means a trend towards contractors taking on more functions than has been usual on the Norwegian continental shelf (NCS) where a greater use of turnkey contracts and integration of services can be seen.

Overall, the chapters in Part I generate a comprehensive picture of the types and dynamics of innovation activities in upstream petroleum. The chapters all point to the close collaborative nature of innovation between suppliers and operators, and also how this has been supported by favourable economic conditions and strong policy support by the Norwegian government. The findings also illustrate that it is a hugely successful model. The institutional support for technology development, particularly in core technologies developed on the NCS, has also been of instrumental importance for the development of the Norwegian technological style, particularly around subsea technologies, which again has been a stepping stone towards international markets (as discussed in Parts II and IV).

The chapters, particularly Chapter 6, also illustrate some of the downsides of this model. Suppliers become strongly dependent on their clients, and have institutionalised a joint responsibility for industry development during downturns. Both Chapters 5 and 6 illustrate how industry downturns function partly as catalysts for innovation, particularly for process innovation, and innovations targeted at curbing costs. The innovation model might, however, be an efficient barrier to more radical transformations – a topic we return to in Part III.

Part II: globalisation of upstream petroleum activities and the role of the supply industry

The chapters in this section have a historical style, based on empirical research from recent decades in Norway (Hanisch and Nerheim, 1992; Ryggvik, 2000, 2013). The main purpose is to provide detailed information of those conditions under which Norwegian companies have developed their international ambitions. With the intensified internationalisation commencing in the early 1990s, the Norwegian supply industry had reached a level where a number of companies were leaders in various sub-segments of the industry. The essential background for their success was the expertise developed in the 1970s and 80s, the first two formative decades, leading to an internationalisation of the industry (see Chapter 2). The main question is therefore how export activities in the supply sector are organised; what types of firm are represented, and the nature of their products.

To illustrate how Norwegian suppliers were able to establish themselves abroad, the chapters use examples from Brazil and the US. It could have been expected that technological developments from the 2000s would continue along broadly similar lines within a continually growing international market. However, if one compares the development in what would remain as the three largest offshore regions for Norwegian suppliers – the North Sea, Brazil and the Gulf of Mexico, there are nevertheless remarkable differences. Some of these differences can be ascribed to different geographical and geological conditions, but others can hardly be explained without including historical and social factors. These differences are expressed both in technological choice and technological style, and the different approaches to organising markets.

Chapter 7 gives an overview of the internationalisation of the Norwegian supplier industry. The chapter concentrates on the supply industry from the early 1990s, with Norway's membership of the European Economic Area (EEA), where Norwegian companies became an integral part of the web of companies that constitute the offshore industries globalised world market.

Chapter 8 focuses upon how the Norwegian supplier industry had become established in a new petroleum region, namely Brazil, what obstacles and opportunities have faced the companies and whether different kinds of institutional arrangements have promoted or prevented the technology transfer process. This chapter also explores the extent to which the local content policy in Brazil affects Norwegian suppliers' strategy entering the deep-sea markets.

Chapter 9 explores the political economy which has shaped the development of major fields in the US Gulf of Mexico. In the light of comparable projects in Brazil and the North Sea, the American sector shows that it has not been only economic responses to geographical and geological conditions, but largely also to social and political conditions which have been decisive for the development of offshore technology. Furthermore, the historical, social and political preconditions which contributed to shaping technological choices offshore in the US help to make clear how other, similar social conditions have affected developments in other offshore markets. These preconditions were also important for many Norwegian suppliers, which, commencing in the early 1990s, worked hard to succeed in the US offshore market

Chapter 10 analyses the growth, current structure and possible future of Norwegian-based employment within the supply industry. It argues that headquarters' functions for rigs and vessels were an important part of the industry even though such highly capital-intensive activity would never be very important for creating onshore employment. Norwegian-based employment will fall with reduced activity on the NCS and only parts of this fall may be compensated by increased exports. As the competencies of both employers (companies) and employees are applicable also in other markets, there are possibilities for gradually more non-petroleum-related employment. To some extent, the chapter differs from the previous three chapters in Part II in the sense that it also discusses the possibility of still maintaining a Norwegian base of the supply industry despite an increasing export orientation and internationalisation.

To summarise the chapters in this part, the main findings are that in the case of Norway, political conditions have affected technological development, not least a strong regulatory regime which emphasised robust technological solutions. Chapters 7 and 10 underline how a Norwegian supply largely was built up by a deliberate government policy where also Norwegian operators were important allies to the governmental actors. At the same time, it might seem that the removal of protectionist barriers and the subsequent internationalisation of the supply industry contributed to detaching the dominant operators from their previous path dependency, and commencing in the 1990s and into the 2000s, a wide spectrum of development solutions were used (see Chapter 2). All these new technologies were important prerequisites to enabling the suppliers to take the step away from the NCS and establish themselves in other petroleum regions such as Brazil and the US. All chapters in this part thus have in common how the Norwegian supply industry, which had been developed and built up under a deliberate protectionist policy, gradually became an international industry without those nationalistic features that had characterised them initially.

Part III: industrial heterogeneity and the implications for future diversification of petroleum supply industry

The three chapters in Part III all address diversification and experiences the petroleum suppliers encountered when moving outside the petroleum market. Diversification is analysed as redeployment of existing resources – knowledge, technological competences, management capabilities, human resources, capital goods – from supply firms into new and related markets. Understanding the heterogeneous industrial basis of petroleum suppliers is therefore relevant for understanding the potential for future development of the supply industry and its potential for diversification to new markets outside oil and gas.

As described above, the knowledge bases required by both operators and suppliers have broadened over time. With increasing technological complexity, knowledge based on a range of scientific and technological fields has been required. The recent history of petroleum extraction and management of the exploration process has been particularly characterised by advanced information and communication technology applications such as 4D seismic surveys, reservoir management and advanced drilling and well operations (Acha and Finch, 2005; Voola, 2006). The technological and scientific knowledge underlying these applications stems from new fields of knowledge such as geophysics, informatics and robotics.

The upstream petroleum industry in Norway developed historically through the transfer of resources from other industries to the oil and gas sector. Large parts of the industry emerged drawing on integrated and transformed knowledge from a range of related and unrelated industries and fields of expertise. The history of the supply industry as such is a story of diversification. Some of the firms in the supply industry therefore predate oil exploration in Norway, and several of the main suppliers maintained activities in several markets throughout the period. Others are actively pursuing alternative futures in the face of a 'post-oil' situation. This has become an important policy issue in recent years, as the joint forces of an industry downturn, increased competition from renewable energy sources and climate change created a policy climate in favour of a 'green shift'.

The studies in this part analyse diversification by existing supply firms into new and related markets. Empirically, each chapter discusses various circumstances that influence diversification efforts by firms, such as types of resources the firms possess, market conditions in different industries, and institutions that support knowledge development and transfer.

Chapter 11 discusses the extent to which the competencies of petroleum supplier firms really are versatile, and what kinds of supply companies are more likely to diversify. Drawing on experience from the past diversification history into oil and gas and existing literature, this chapter examines key assumptions about diversification behaviour by oil and gas supply firms. This shows that most firms in the petroleum supply industry are diversified and involved in other markets, but mainly in three related sectors: shipping and the maritime sector, land-based manufacturing, and construction. The main characteristics of firms diversifying into multiple sectors are that they are product-oriented, small, young, not very affluent, but strong in generic competences.

Chapter 12 examines strategies taken by so-called early diversification among supply firms, i.e. firms that started to explore alternative market opportunities outside petroleum at an early stage. The challenges involved in entering new markets are considered, particularly how the unique collaborative innovation model in oil and gas to which the suppliers are accustomed also represents a fundamental challenge for succeeding in other markets, even when technologies are related.

Chapter 13 addresses this last issue in detail through a case study of diversification of petroleum firms into offshore wind power. A description is given of the intermittent pattern of diversification by petroleum-related companies, largely in correspondence with price fluctuations in crude oil. This entails that petroleum-related companies diversify when their core market suffers and jump back again when times are good. But even though this seems like a straightforward pattern, a detailed case study of several firms shows that investment of resources specific for new niches determines whether the engagement pattern is intermittent or steady over time.

A general finding from Part III is that diversification processes are demanding and challenging. The processes involve access to, and use of, wide and advance knowledge, technologies and other resources, which may be deployed and invested in other industrial sectors. In addition, and probably most important, are non-technological factors, particularly relations with users and costumers. Characteristics of contracts, funding and trust create a specific collaborative and interdependent relationship between supply firms and oil companies that is distinctively different from other sectors. As most of the supply industry today is either only selling to the petroleum market and two or three other sectors, it is argued that there is a potential risk for future diversification which public policies should address to achieve a 'green transformation'.

Part IV: perspectives on economic development of oil economies

The last part of the book, Part IV, provides several reflections on the development of the Norwegian petroleum sector and its long-term influence on national economic and industrial development as well as the role of the supply industry in the upstream petroleum industry in an international perspective.

As mentioned above (see also Chapter 3), there is a long-lasting discourse on the role of natural resource industries for long-term economic development. Many petroleum economies have experienced the 'resource curse' and the 'paradox of plenty' with low growth rates, de-industrialisation and political instability. Compared to many emerging oil economies, Norway is seen as the exception, and often regarded as a model to be followed. The development of a strong petroleum supply industry plays an important part in the model. The local content strategies and policies to promote the development of local firms and production capacity linked to natural resource industries are introduced in other oil economies (see Chapter 8). Norway succeeded well, making services and products for the international petroleum sector a specialisation. Norway's experiences raise the issue concerning the extent to which natural resources form the basis for long-term economic development (Ville and Wicken, 2012; Andersen *et al.*, 2015).

The Norwegian supply industry emerged, expanded and became international during a period of transformation of the global industrial structure of the petroleum sector. Until the 1980s large integrated oil companies developed technologies which they deployed in different petroleum regions. During recent decades this changed, and specialised suppliers have gradually played a more central role for innovation and technological development in the industry (see section 1.2).

The implication was a change of relations between supply firms and oil companies characterised by close collaboration; the development of the collaborative innovation system was strongly influenced by public policy (see Chapter 2).

Chapter 14 presents the Norwegian oil experience from a policy perspective, facing the challenges of developing an independent oil industry. It describes various policies to promote a high level of local content, infant industry protection and policies to develop a knowledge-intensive supply sector, and in this way avoid pitfalls such as de-industrialisation and 'Dutch disease'; also, governments' intentions to protect the domestic economy from fluctuations in the oil price and earnings resulting in the world's largest sovereign wealth fund. In spite of this, it is argued that the large financial assets of the fund may result in a consumption-driven economy with low motivation for structural change.

Chapter 15 provides an international and cross-industry perspective on the Norwegian oil and gas industry. The overall perspective discussed is how unique the petroleum innovation model is, or whether it is part of a broader industry trend towards disintegration and collaboration. In this chapter, it is argued that oil and gas shares many governance features with other sectors, outside the natural resources sector, in which producers confront intractable technological and market uncertainty. The chapter outlines the character of vertical disintegration and collaboration in the oil, pharmaceuticals and automobile industries and explores the distinctive and remarkably formal form of 'contracting for innovation' that organises innovation in each. By highlighting the common ways in which these industries systematically generate possibility through collaboration, the chapter destabilises the conventional distinction central to development economics, between natural resources and industry in the development process.

Conclusion: successful natural resource industry and economy

There is general agreement that Norway has succeeded well economically as a natural resource-based or oil economy. Norway is, however, not the only small or medium sized economy specialising in natural resources with strong economic performance and competiveness over the latest decades. The same has been the case for Denmark, Finland, Sweden, Iceland, Australia, New Zealand and Canada (Smith, 2007).

Compared to many emerging oil economies today Norway had an advantageous position from the start. It was a developed industrial economy. Many industries relevant for offshore oil production – shipping, shipbuilding, construction, fisheries, ICT, finance and other – became gradually involved in the petroleum sector (see Chapters 2, 10 and 14). A number of the firms were operating in international markets, and some (like shipping) had worked with oil companies over long periods of time. Well established local firms could from the beginning draw on their competence from earlier production and international experience to enter the global value chain of the petroleum sector.

How did the petroleum supply industry become an important source for innovation and dynamics in the Norwegian economy – and also become a large export industry? In this book we have analysed Norway's development drawing on approaches from innovation studies which use knowledge as the most important resource and learning or innovation as the most important process for long-term economic development (Lundvall and Johnson, 1994). We have shown that Norway during the oil era has built up strong and heterogeneous knowledge bases relevant for the petroleum sector. The supply industry constitutes the major knowledge source for innovation in the upstream petroleum sector. Together with strong R&D institutions they have enabled innovation and contributed to the transformation of Norway into a natural-resource-based knowledge economy. The result has been an economy with – 'double specialisation' in oil and gas: petroleum products constitute by far the largest industry and export sector, technology and services for the petroleum sector is the second largest. The rapid growth of production of goods and services for the oil and gas market is the main factor behind the relatively low level of de-industrialisation. The growth of the petroleum supply industry compensated for decreased production in other manufacturing industries.

Public policies played an important role in developing resources for innovation in the petroleum sector from an early period by introducing local content policies and infant industry protection. From the late 1970s new policies to transfer knowledge and expertise to the supply industry and national research institutions were introduced, supporting the build-up of relevant R&D capacity in the economy. The intention was to give Norwegian firms access to new technological fields to push foreign oil firms to transfer technology and expertise to Norwegian firms and organisations (see Chapter 14). The result was a rapid increase in industrial R&D in the industry, and from the turn of the millennium direct public investment in R&D further supported the capabilities for innovation (see Chapters 2 and 3).

A central topic of the book is the importance of a collaborative mode of innovation for the long-term development of the upstream petroleum industry. The importance of collaboration between oil companies, supply firms, R&D institutions, and public agencies is illustrated at local (Chapter 4), national (Chapters 2 and 3) and international level (Chapters 7 and 8). This model emerged through processes involving public policy in cooperation with oil companies and the supply industry from the 1990s (see Chapter 2). A central aspect is that this mode engages also non-scientific expertise and knowledge in innovation processes. The combination of scientific and experience-based knowledge has resulted in efficient technological and organisational solutions. From this perspective we may argue that Norway's successful development as a petroleum economy is based on the combined development of strong and relevant knowledge capabilities and a collaborative mode of innovation which has promoted effective use of the existing innovation capabilities. In Chapter 15 it is argued that this mode of innovation has become dominant in the global value chains of many industries, and that competence in this mode of innovation may be seen as a generic competitive advantage for succeeding in international competition.

The importance of innovation capabilities and efficient models and institutions for innovation processes will remain central for petroleum industries and economies in the future. The industry faces new types of challenges demanding new types of solutions; from competition from new ways of producing oil and gas, new forms of renewable energy and climate policies. These challenges may demand new types of knowledge bases and new modes of innovation, demanding transformation processes of a similar type as the industry experienced during the 1970s and again during the twenty-first century. As was the case during the start-up of the petroleum industry, future development will have to build on the existing industrial capability and the supply industry will have an important role to play in this transformation.

References

Acha, V. L. (2002). *Framing the past and future: the development and deployment of technological capabilities by the oil majors in the upstream petroleum industry*. PhD thesis, University of Sussex.

Acha, V. and Finch, J. (2005). Paths to deepwater in the international upstream petroleum industry. Green, K., Miozzo, M. and Dewick, P. (eds.) *Technology, Knowledge and the Firm: Implications for Strategy and Industrial Change*. Edward Elgar: Cheltenham, UK, 73–91.

Andersen, A. D. (2012). Towards a new approach to natural resources and development: the role of learning, innovation and linkage dynamics. *International Journal of Technological Learning, Innovation and Development*, 5(3), 291–324.

Andersen, A. D., Johnson, B. H., Marín, A., Kaplan, D., Stubrin, L., Lundvall, B.-Å., and Kaplinsky. R. (2015). *Natural resources, innovation and development*. University of Aarhus.

Bagheri, S. K. and Di Minin, A. (2015). The changing competitive landscape of the global upstream petroleum industry. *The Journal of World Energy Law & Business*, 8(1), 1–19.

Beyazay-Odemis, B. (2016). *The Nature of the Firm in the Oil Industry*. New York: Routledge.

Bridge. G. and Wood, A. (2005). Geographies of knowledge, practices of globalization: Learning from the oil exploration and production industry. *Area*, 37(2), 199–208.

Corden, W. M. and Neary, J. P. (1982). Booming Sector and De-Industrialisation in a Small Open Economy. *The Economic Journal*, 92(368), 825–848.

David, P. and Wright, G. (1997). Increasing Returns and the Genesis of American Resource Abundance. *Industrial & Corporate Change*, 6(2), 203–245.

Engen, O. A. (2009). The development of the Norwegian Petroleum Innovation System: A historical overview. Fagerberg, Mowrey and Nelson (eds.), *The Oxford Handbook of Innovation*. Oxford: Oxford University Press, 179–207.

Gylfason, T. (2001). Natural resources, education and economic development. *European Economic Review*, 45(4), 847–859.

Gylfason, T. (2004). Natural Resources and Economic Growth: From Dependence to Diversification. Broadman, H. G., Paas, T. and Welfens, P. J. J. (eds.), *Economic Liberalization and Integration Policy*, Springer, 201–231.

Hanisch, T. and Nerheim, G. (1992). *Norsk Oljehistorie. Fra vantro til overmot? [The History of Norwegian Oil: From Disbelief to Arrogance?]* Oslo: Leseselskapet.

Inkpen, A. C. and Moffett, M. H. (2011). *The global oil & gas industry: Management, strategy & finance*. Houston, US: PennWell Books.

Lundvall, B.-A. and Johnson, B. (1994). The Learning Economy. *Journal of Industry Studies*, 1(2), 23–42.

Mahroum, S. and Al-Saleh, Y. (eds.). (2016). *Economic Diversification Policies in Natural Resource Rich Economies*. Abingdon: Routledge.

Morris, M., Kaplinsky, R., and Kaplan, D. (2012). One thing leads to another — Commodities, linkages and industrial development. *Resources Policy*, 37(4), 408–416.

Nelson, R. and Sampat, B. (2003). Making sense of institutions as a factor shaping economic performance. *Journal of Economic Behavior and Organization*, 44, 31–54.

Perrons, R. K. (2014). How innovation and R&D happen in the upstream oil & gas industry: Insights from a global survey. *Journal of Petroleum Science and Engineering*, 124, 301–312.

Pinder, D. (2001). Offshore oil and gas: global resource knowledge and technological change. *Ocean and coastal management*, 44, 579–600.

Ryggvik, H. (2000). *Norsk oljevirksomhet mellom det nasjonale og det internasjonale. En studie av selskapsstruktur og internasjonalisering.* [Norwegian oil business between the national and international. A study of corporate structure and internationalization.] PhD thesis. University of Oslo.

Ryggvik, H. (2013). *Building a skilled national offshore oil industry. The Norwegian experience*. The Confederation of Norwegian Enterprise (NHO), Oslo.

Ryggvik, H. (2017). *The offshore oil industry in Brazil: A Norwegian historical perspective.* Unpublished manuscript.

Rystad Energy, (2017). *Internasjonal omsetning fra norske oljeselskaper* [International sales from Norwegian oil companies], report to Minstry of Petroleum and Energy, 31. October 2017, Oslo.

Sachs, J. D. and Warner, A. M. (1995). *Natural resource abundance and economic growth.* National Bureau of Economic Research Working Paper, no. 5398, December.

Sachs, J. D. and Warner, A. M. (2001). The curse of natural resources. *European Economic Review*, 45(4–6), 827–838.

Shuen, A., Feiler, P. F., and Teece, D. J. (2014). Dynamic capabilities in the upstream oil and gas sector: Managing next generation competition. *Energy Strategy Reviews*, 3, 5–13.

Smith, K. H. (2007). *Innovation and growth in resource-based economies*, CEDA Growth 58: Competing From Australia, Committee for Economic Development of Australia, Melbourne, 58 (2007).

Ville, S. and Wicken, O. (2012). The Dynamics of Resource-Based Economic Development: Evidence from Australia and Norway. *Industrial and Corporate Change*, 22(5), 1341–1371.

Von Tunzelmann, N. and Acha, V. (2005). Innovation in 'low-tech' industries. Fagerberg, Mowrey and Nelson (eds.), *The Oxford Handbook of Innovation.* Oxford: Oxford University Press.

Voola, J. J. (2006). Technological change and industry structure: A case study of the petroleum industry. *Economics of Innovation and New Technology*, 15(03), 271–288.

Wicken, O. (2016). Industrial diversification processes and strategies in an oil economy: Norway. Mahroum, S. and Al-Saleh (eds.), *Economic Diversification Policies and Natural Resources*, Routledge Explorations in Environmental Economics.

Additional sources

U.S Energy Information Administration. (2018). Spot Prices for Crude Oil and Petroleum Products: Europe Brent Spot Price FOB. Retrieved 02.2018.

Statistics Norway. Statistikkbanken, National accounts: Gross fixed capital formation and capital stocks, by type or industry 1971–2017. Retrieved 06.2018.

Part I

2 The evolving sectoral innovation system for upstream oil and gas in Norway

Ole Andreas Engen, Erlend Osland Simensen and Taran Thune

Introduction

During the first decades of the twenty-first century, significant changes in the Norwegian petroleum sector have occurred. Increased international integration, changes in the composition of actors and an increasing amount of resources spent on R&D and development of new technologies established the Norwegian petroleum economy among the most prosperous in the world. There has been a high degree of collaboration between suppliers of technological knowledge and new solutions on the one hand, and advanced users among the oil companies on the other. This gave the innovation system momentum that opened up for several new and advanced technical solutions necessary in a 'maturing' petroleum region, and based on this increased exports of Norwegian technologies to other petroleum markets.

This chapter examines the development of the innovation system for upstream petroleum in Norway over the last two decades. To support this analysis, we draw upon the concept of the sectoral innovation system (Malerba, 2002), and in particular Malerba's perspective on the dynamics of sectors (Malerba, 2005; 2007). The empirical part of the chapter outlines the main developments of the innovation system that have occurred over the last 18 years, looking at development in terms of changing actors, technologies and institutions. We conclude the chapter by discussing the co-evolutionary nature of sectoral innovation system changes and the overall innovation system transformation of the Norwegian upstream petroleum sector.

Sectoral innovation system and system transformation

The concept of the sectoral innovation system was proposed by Malerba in 2002 to account for the industry-specific nature of innovation. Along with many other scholars, Malerba argues that industrial sectors have specific patterns of technological development and deployment, referred to as technological regimes (Dosi, 1982; Nelson and Winter, 1982). According to Malerba, technological regimes influence innovation in a sector or an industry by setting boundaries and defining trajectories for problem-solving activities (Malerba, 2005). These regimes influence both the rate and kind of innovative activities performed by firms as well as their organisation. Malerba (2002, p. 65) claims that technological regimes are more or less unique to industrial sectors:

> Heterogeneous firms facing similar technologies, searching around similar knowledge bases, undertaking similar production activities and embedded in the same institutional setting, share some common behavioural and organisational traits and develop a similar range of learning patterns, behaviours and organisational forms.

The following three elements and processes are key elements of sectoral innovation systems:

- Actors at different levels of aggregation, and the interaction between them.
- Technologies and the knowledge base and learning processes that support the development and deployment of technologies.
- Institutional framework conditions that shape innovative efforts, such as norms, routines, established practices, standards, laws and policies.

The sectoral innovation systems perspective has been partly used to classify sectors and to discuss how sectors differ in innovation activities. In the classic formulation of the sectoral innovation system, sectors comprise firms that produce and sell similar products (Malerba, 2002). Accordingly, we define the petroleum supply industry as an industrial sector that consists of suppliers who develop and sell their products and services to a limited number of petroleum companies. Operators (oil companies) largely influence the demand side and the selection of innovations, and consequently the direction of innovation activities. In addition to this, the public sector, and particularly governmental regulation and policies, play a significant role.

Characterising and describing different sectors is just one role for sectoral systems analysis. It is equally important to understand how industrial sectors evolve, and the role innovation plays in the evolution of sectors (Malerba, 2006). Thus, it is necessary to understand fundamental processes of change and also which factors give shape to sectoral evolution. Malerba discerns three driving forces of sectoral change. One is change in demand by customers and users; the two others are changing institutional framework conditions and changes in the supply of technologies.

A main point for Malerba is that sectoral change should be understood as co-evolution among elements that make up the system, and not as driven by a single element. He argues that in order to study sectoral change, one has to study the 'coupled dynamics' between changes in actor composition, knowledge and technology, institutional conditions and demand conditions in a longitudinal fashion to trace whether the 'system variables' change together, and also to discern potential feed-back loops between elements of the system.

With these perspectives in mind, this chapter describes changes in demand conditions, in the actor composition, technological solutions and institutional framework conditions in the Norwegian petroleum sector during the last two decades.

Transformation of the sectoral innovation system for upstream petroleum in Norway

When oil was first discovered off the coast of southern Norway in the 1960s, Norway had no experience in petroleum production. The initial phase of a Norwegian petroleum innovation system in the sixties and seventies was thus characterised by collaboration between multinational oil corporations and an emerging domestic industry, and technology communities with the aim of developing absorptive capacity in receiving and using new technology. In this process, scientific and industrial competences in the maritime and shipping industry, and later the mechanical and defence industries, were transformed into becoming parts of the petroleum supply industry (Engen, 2009).

As elaborated in Chapter 14 in this book, the national capacity for governing its petroleum resources matured alongside public policy, industrial development and technologies. The development of operator companies with Statoil and Hydro as dominating actors, and the development of a large and complex supply industry, signified that petroleum activities in Norway were entering a new phase in the late seventies and eighties. In the new phase, resources were invested in education, research and innovation, and the innovation system even in this initial stage was characterised by a high degree of interaction between users and suppliers of knowledge and technology. The capacity to develop and deploy advanced technologies in collaborative networks was also supported by a benevolent policy and tax regime.

In the first half of the 1990s, all actors in the Norwegian oil industry realised that it was necessary to break down the technological style associated with large and cost demanding gravity platforms and complex development projects organised by the operators (see later section). Innovation processes within seismic, drilling, production installations, subsea and contractual relations had thus been going on for a while when we reached the end of the decade (Engen, 2002). These initial developments are important to have as a backdrop, when we now turn to the developments occurring in the mature phase of petroleum production in Norway after 2000.

Development of actor constellations

By 2000, Norway had become an advanced petroleum-exporting nation where Statoil had been privatised and later became the dominant player on the Norwegian shelf through mergers and acquisitions. International operators became less important, and a range of large suppliers and oil service companies (both domestic and international) became key technology suppliers and partners, particularly in the development of new fields and upgrading of older fields. The experience of these actors acquired on the Norwegian domestic shelf with its advanced technological requirement also meant that the export of technologies to new petroleum regions became an increasingly important activity (see Chapter 7).

From 2000 to 2017, the innovation system changed substantially through the enrolment of a range of new actors on the Norwegian shelf. In this period a multitude of large and small suppliers and service firms emerged, and these firms developed several technology niches and specialised clusters. Moreover, new oil companies emerged and some of the large international companies started to withdraw from the North Sea area. This was a co-evolutionary and interconnected process in the sense that the new composition of actors directly mirrored a deliberate policy of the Norwegian authorities, i.e. to maintain the activity level on the shelf, which again derived from changes in external (natural and financial) conditions. It was also a reflection of an international tendency towards disintegration of large petroleum companies and the creation of complex value chains in the global petroleum industry (Acha, 2002).

Maturation of the Norwegian continental shelf

An important factor for changes in actor constellations and demand conditions was that central actors in the innovation system realised already in the 1990s that the Norwegian shelf had reached a 'stage of maturity' (White Paper 26, 1993–1994). A 'stage of maturity' refers to several factors. First, that the discoveries of the large oilfields, the era of the so-called 'elephants', e.g. Ekofisk, Frigg, Statfjord and Gullfaks, was over. Second, and related to the first, that future field discoveries were most likely small fields connected to existing infrastructure – so-called satellites, small independent fields or large field-structures consisting of both oil and gas, and thus difficult and costly to develop. Third, that most expected field discoveries were located in deep water. The new demand conditions stemming from the natural conditions of the Norwegian continental shelf required dramatic cost reductions along the entire value-chain to make it economically feasible to develop new and less profitable fields, but also increasing investments in new technologies was necessary (OG21, 2016).

This development also impacted the constellation of actors in the sector since a maturing shelf requires increased 'tail production'. Tail production refers to petroleum fields where the cost of extracting the oil will be higher than its market value unless major improvements in technology and work processes are made. Thus, large actors decide to withdraw, sell out or establish new companies. This implies that small independent companies which are specialised in this form of production take over and contribute to restructuring the market. This also occurred on the Norwegian continental shelf. The newcomers (e.g. Lundin, Wintershall Noble Centrica, Det Norske Oljeselskap) represented diversity concerning both how they were organised and the role they played on the shelf (Norwegian Petroleum Directorate, 2017). The new oil companies were mostly involved in the exploration phase. Some of the new oil companies, however, chose to operate in all phases, i.e. from exploration to development and operation.

New oil companies and growth in the supply industry

In January 2017, there were 27 active operators and 26 partners holding production licences. This number has not changed significantly during recent years. Changes among operating companies have mainly been related to mergers, acquisitions and bankruptcies among contractors and suppliers. Some of the major oil companies have, however, shown increased interest in the Norwegian continental shelf, and major companies such as ConocoPhillips and Shell have received more allocations of blocks compared to previous years. At the same time, a number of major international companies have changed their strategy for presence on the Norwegian continental shelf and withdrawn as licensed operators (e.g. ExxonMobil, BP). Although the composition of actors on the operator side has become more diverse after the turn of the millennium, Statoil is increasingly the dominant player. By 2016, Statoil's share of production as a licensee was 31.4 per cent, and as operator 67.1 per cent of total production (Norwegian Petroleum Directorate, 2016). Through its dominant position, Statoil has a significant impact on the technological development in the industry, and is crucial for the overall business on the shelf.

Alongside the developments on the operator side of the sector, significant developments on the supply side have also occurred in recent decades where there has been a substantial increase in the number and variety of domestic petroleum supply firms. The supply industry is notoriously difficult to capture in statistical exercises, thus it is difficult to present precise measures about the growth of companies and employees in this part of the industry. Reports indicate a growth from about 1,500 petroleum-related firms (firms that sell products and services mainly in the petroleum market) in the mid-2000s (Vatne, 2013), to around 3,000 firms in 2014 (Blomgren *et al.*, 2015). Similarly, around 100,000 people were employed in the petroleum sector (including suppliers) in 2007, and around 300,000 people at the end of 2014. Of these, about 13 per cent were employed by the operator companies or on the offshore installations, while the rest worked in supply and service companies. A total of 43 per cent of the employees – around 140,000 – worked in firms that were direct suppliers to the petroleum industry at the end of 2014. In total, 11 per cent of all private sector employment in Norway at the time was directly or indirectly connected to petroleum operations in Norway or abroad (see Chapter 10).

These figures clearly indicate that the petroleum supply sector is a significant part of the Norwegian economy and has grown substantially during the last two decades. Moreover, the increased growth in firms has also been followed by increased specialisation into multiple niches or sub-markets, and the development of a complete value-chain model connected to domestic petroleum operations. The different segments of the supply value-chain tend to be regionally clustered, and this again has given rise to the development of several petroleum supplier clusters in Norway (Rystad Energy, 2015).

After 2000, several specialised clusters developed along the Norwegian coast and outside the capital region of Oslo (Chapter 4). The companies were located in geographically defined areas and to some extent within the same types of goods and services. In the Oslo region, a well-established engineering environment developed with a wide range of knowledge-based services and a concentration of seismic companies, while the Bergen region became the centre for the maintenance of platforms and underwater equipment. In Kongsberg, companies which previously specialised in defence technology evolved into subsea technology including automation and dynamic positioning equipment. Southern Norway contributed with world-leading companies in drilling technology while on the northwest coast of Norway a maritime cluster developed representing a complete shipbuilding and ship equipment network for advanced offshore vessels, among other things. Stavanger maintained its role as a stronghold for companies offering services and modifications to existing platforms and equipment, and therefore has the largest employment share of the industry (see Chapter 4).

As this part of the chapter has illustrated, the last two decades have witnessed a maturation of the Norwegian continental shelf. This implied changes in the constellations of actors with three parallel trends – a concentration of resources around one dominant domestic petroleum company, withdrawal of international oil companies, and the growth of new and independent firms where some specialised in tail production. In parallel with the increased nationalisation of resources, a strong growth in the number and kinds of supply firms has occurred, many of which offer technologies and knowledge within a range of specialised niches to operators and particularly to Statoil. In the next section, we look at the development of the innovation system from the perspectives of technologies and the particular demand conditions that have shaped a new 'technological style' in Norwegian petroleum production during the last two decades.

Development and deployment of technologies

New demand conditions on the maturing shelf

With the maturing of the shelf, the time for technology development was not over around the turn of the century. In order to reach the increasingly inaccessible oil and gas deposits remaining in Norwegian territory, suppliers and oil companies in collaboration have developed and implemented a range of novel technologies during the last two decades. The state of the shelf increased the need for prolonging activities in existing oil wells as well as the need to explore whether deposits previously not seen as technologically or economically viable now were recoverable. 'Enhanced oil recovery' therefore became a prioritised task in this era of the Norwegian oil activities.

The demand for enhanced recovery has fuelled technological development, particularly in two domains – subsea technology and reservoir knowledge. Subsea technology is not a single technology but consists of a wide array of solutions that need to be integrated into a system. In other words, 'subsea' is a generic term for submerged petroleum facilities often installed on the seabed, and includes components such as Pipeline End Terminations (PLETs), manifolds, umbilicals, risers and floating vessels. As more subsea projects have been developed and installed on the Norwegian shelf, the demand for development of more sophisticated components in this system has increased. With more components installed on the seabed, oil and gas wells could be managed by temporary rather than permanent installations such as Floating Production, Storage and Offloading (FPSOs).

Oilfield developers have also achieved better recovery rates through improved design (engineering) and surveying of the oilfield, as well as improved ability to develop geographically more remote oil wells through subsea systems. An important part of the increased knowledge about the reservoirs is the development of advanced seismic surveys of new and existing fields. Particularly the development of time-lapsed 3D models – so-called 4D – has been important to the success rates in drilling. This has enabled oil companies to drill more precisely and in addition to reducing exploration costs; it has made the operators willing to drill more marginal fields since the overall risk has decreased (Fjose *et al.*, 2014). Subsea development and better seismic understanding of fields is thus connected through the demand for enhanced oil recovery and the possibility to exploit less accessible resources.

Another technology domain related to the demand for enhanced oil recovery, and which has received significant policy attention in recent years, is the possibility for offshore carbon capture and storage. Injection of seawater, and more lately CO_2, into the reservoirs is now also being used in offshore projects. This procedure stems from land-based oil wells, and was first applied in the US. Whereas this leads to increased oil recovery by 'pushing' the gas out of the reservoirs, it also allows storing of carbon dioxide beneath the sea level.

New technologies deployed after 2000

With these new demand conditions and the new technologies in mind, we have summarised important steps in the deployment of new technologies on the Norwegian continental shelf in the period from 1999 to 2017. These are developments around two technology domains; subsea, and increased reservoir monitoring by the use of seismic technologies. In Figure 2.1, the development and deployment of these technologies on the NCS is illustrated.

In 1999, the Åsgard field was the first to use FPSO technology on the Norwegian shelf. Operators chose floating vessels instead of platforms due to their mobility and lower costs in setting up, making smaller and more remote fields economically viable. They are also practical for gas fields due to their ability to process the gas immediately, reducing the need for expensive, long-distance

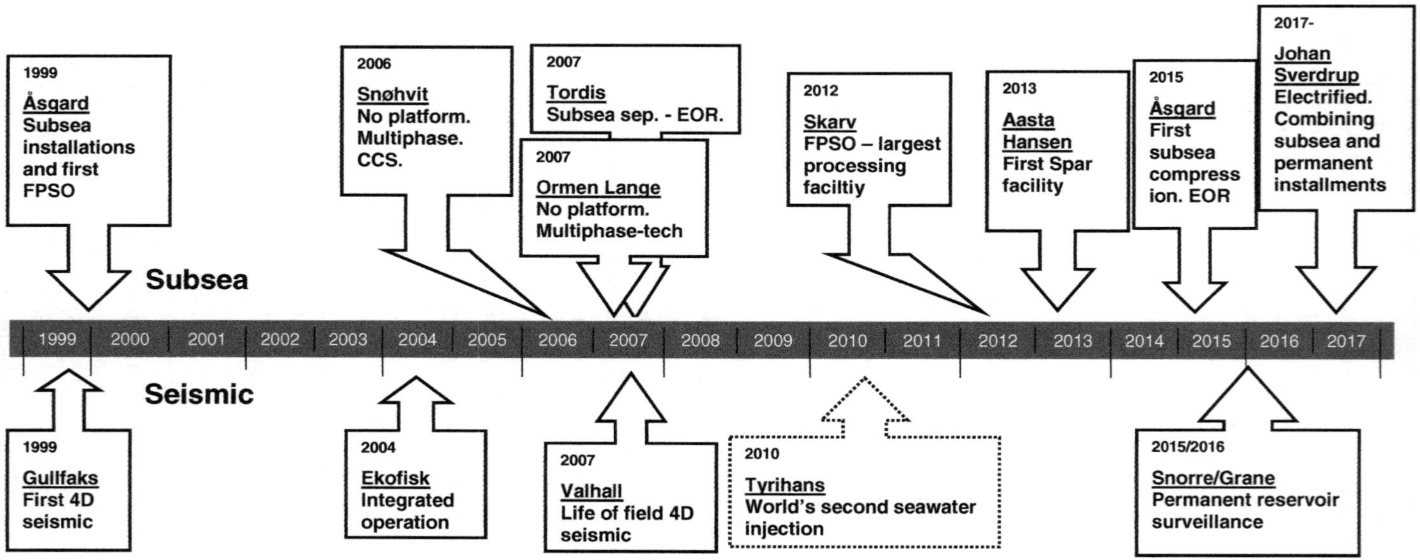

Figure 2.1 Main petroleum technologies developed and deployed on the NCS after 2000.

pipelines. Much later, in 2012, Aker further developed FPSO technology on the Skarv field. Here, they built one of the largest offshore gas processing plants that was anchored to the seabed and operated solely with floating vessels. The gas from Skarv connects directly to the European gas market through the Gassled distribution pipelines.

The multiphase technology developed in Norway in the 1980s has been one of the most important enablers of full-scale subsea systems, since it allows the transport of unprocessed masses over long distances. Snøhvit (2006) is a good example of a field that has utilised this technology. This field is located far north in the Barents Sea and consists predominantly of natural gas resources. It was the first field with neither a fixed nor a floating unit. Instead, the gas goes through 140 km multiphase pipelines directly from the seabed-mounted units to the Liquefied Natural Gas plant on land. Further south, Ormen Lange used the same technology with no platforms and long multiphase pipelines. This was also Norway's first deep-water project (850–1,100 meters).

Additionally, the implementation of new subsea technologies further allows for upgrading of older fields. For example, to increase the recovery rate when the reservoir pressure drops as the field matures, subsea units can prolong the field by installing compression units on the seabed. The Tordis field is the first example of this; FMC Technologies upgraded it with subsea units in 2007, and it was regarded as the first full-scale commercial subsea separation, boosting and injection system. This has also been carried out on the Åsgard field, which was upgraded with the world's first subsea compression unit developed by Aker Solutions in 2015.

Several fields have contributed to increased knowledge in the seismic and reservoir knowledge technology domains. Gullfaks, an old and complex field in the North Sea, was the first to introduce a time-lapse analysis of the field in 1999. 4D analysis, as it is also called, is 3D seismic analysis that is being performed over time to observe the development of the oil reservoir. This made it possible to drill two additional wells that were interpreted as being undrained based on the repeated seismic data. To further increase the reservoir knowledge digitally, Ekofisk installed the first onshore centre in 2005. In these centres, all operations are computer simulated, significantly reducing costs by requiring fewer offshore operating days. These type of operating centres are now industry standard.

The development of the large field, Johan Sverdrup, shows how the implementation of a range of new solutions has led to new possibilities for field development on a maturing shelf. Production will commence in 2019 and the field will include many of the new solutions outlined in this chapter. With an ambition of a 70 per cent recovery rate and an extraction horizon of 50 years, technology has been in focus from the beginning. The plan is to apply a combination of floating production and drilling platforms, subsea production and factories, long-distance pipelines, shuttle tankers, as well as fixed production platforms. With such an immense range of components working together and the vast dimension of the field, the digital simulation software requirements and organisational capabilities are probably the most radical innovations.

In that sense, the Johan Sverdrup field is a prime example of innovation in the oil and gas sector. The technologies may not be new individually, but they have never been installed in such an environment, nor implemented together in such a combination. Statoil and suppliers are also in the process of developing production designs for less profitable fields such as Johan Castberg through the use of advanced technologies, but also with a strong focus on system integration and process innovations to fit new market and environmental conditions.

Between 2000 and 2017 we have seen that the development of seismic- and subsea-related technologies have led to increased oil recovery and further development of previously unprofitable fields. This development has also resulted in more efficient operations of complex and remote fields. The important drivers behind this development seem to be the general increase in seismic technology, horizontal well drilling and data processing, but also Norwegian 'inventions' such as multiphase technology. It is therefore reasonable to claim that the broad range of deep sea and underwater technologies represent a new technological style significantly different from the technological style of the eighties and nineties.

Changing institutional framework conditions

Developments of the constellations of actors and technologies to fit petroleum production on a maturing shelf, as already indicated, has also been supported by institutional changes. However, what is important to have in mind is that compared to most other sectors in the Norwegian economy, development of the petroleum industry is a highly political exercise, or stated differently, it has been significantly shaped by a strong policy willingness to support the operators (partly state owned), but also to develop a large national supply industry. In fact, there is probably no other industrial sector in Norway that has been shaped by public policies to the same extent as the petroleum sector.

In this part of the chapter, we discuss some examples of recent policies and institutional arrangements that have supported the sector's development in the maturation stage. When policy makers and industrial actors together realised in the late 1990s that external financial and environmental conditions could affect future investment on the Norwegian continental shelf, they activated a large number of institutional tools to stimulate increased investment and activities among operators and suppliers.

Institutional support for increased activity and variety of actors on the shelf

The concession system is the most important instrument for regulating the activity level. Another accompanying instrument is the different kinds of incentives built in the taxation system. Formal concession laws have been more or less unchanged since 1965, but the practice has been adjusted over the years mainly due to changing political objectives. In order to maintain the activity level after the 1990s, two policy measures associated with concession practise were implemented: the introduction of pre-qualification of new operators and licensees in 2000, and second, the establishment of the APA-system (allocation in predefined areas) in 2003, which was intended to stimulate increased activity in areas where the geology was relatively well-known. Another measure associated with the tax system was the so-called reimbursement system for exploration costs, implemented in 2005. The petroleum taxation system is based on the rules for ordinary company taxation but because of extraordinary returns (resource rent), the oil companies are subject to additional special taxes, depreciations and deductions. The reimbursement system came in addition to prevailing petroleum tax rules in order to decrease the entry barriers for new actors and encourage economically viable exploration activity.

All these formal institutional changes represented policy actions with the intention of delaying any decline in the activities, focusing on the operator side of the industry. Other policies were intended to support the development of Norwegian suppliers. One important institutional tool in this regard is a string of programmes implemented from the mid-1990s with the aim of increasing the competitiveness and efficiency of petroleum production in Norway.[1] The first of these was the so-called NORSOK programme. The prime goal of NORSOK was to support technological development, development of joint technological standards, and cost efficient organisational developments (Engen, 2002; Ryggvik, 2013).

In 2001, KONKRAFT replaced NORSOK as an arena for joint co-operation in order to work for cost efficiency and competitive abilities. Another organisation, INTSOK, set up jointly by the government, operators and suppliers, became an important incubator for the supply industry concerning its international establishment. In general, inter-organisational collaborative organisations with clear policy strategies have continued to play a significant role in connecting industrial actors and the political system and constitute an institutional force for business strategy, technological development and internationalisation. This has also extended into the technology development area where a collaborative model of setting a joint R&D and innovation agenda – the so-called national technology strategy process for oil and gas or OG21 – was first implemented in 2001.

One result of the NORSOK process was that Engineering Procurement, Construction and Installation (EPCI) contracts become the usual type of contract on field development projects carried out in Norway after the mid-1990s. These contracts imply that the operator gives the entire responsibility for a development project involving new fields to a single main supplier. The main suppliers thus have the responsibility for selecting the equipment suppliers. The EPCI contracts increased both the opportunities and the risk for the main suppliers, but they did not guarantee a Norwegian share of the contracts. However, high technological competence among the sub-contractors resulted in a Norwegian value creation of approximately 50 per cent for most of the projects carried out between 2000 and 2015, even though the construction phase of some projects was located in Korea (OED, 2015). In this case 'value creation' refers to the share of Norwegian technology suppliers in the projects. Another factor explaining the success of the Norwegian sub-contractors is the use of technical standards.

Another legacy from the NORSOK programme was the implementation of so-called NORSOK standards. Standards can be understood as common protocols that meet the intentions of the regulators and operators' requirements – i.e. they define which solutions qualify as acceptable technical and organisational fulfilments of such requirements. The NORSOK standards have been developed by the Norwegian petroleum industry and according to its owners, express best practice and 'competitiveness both nationally and internationally' (NOROG, 2016). The standards are a prerequisite for qualifying as a supplier on the Norwegian shelf. In accordance with the standards, the suppliers (or in some cases clusters of suppliers) normally put together packages that fulfil the contract and often guarantee that the production and deliveries take place in Norway. The requirement of using NORSOK standards in the EPCI contracts partly resulted in increased activities on the shelf in the period 2000–2017 and were channelled through the Norwegian-based supply industry.

Institutional support for technology development and deployment

As seen above, a range of new technologies has been developed and deployed on new fields during the last two decades. This increasing 'technological turn' has to be seen in light of the maturing shelf, available funding due to relatively high prices for crude oil, but has also been greatly supported by public policy. The latter has been achieved through the development of a broad portfolio of innovation and R&D policies and programmes, targeting firms in different parts of the petroleum sector and public research organisations and, not least, the interactions between them.

The basic premise for innovation and R&D support in the petroleum sector is that the petroleum industry carries out and finances the lion's share of R&D and that such activities are supported by a benevolent petroleum tax regime. Through this system, 50 per cent of operators' R&D expenditures are tax deductible. For the oil companies that are licensed owners and operators of fields, this specific tax policy is the most significant support scheme.

The main changes in the public innovation support system from the late 1990s are, first, the development of a range of targeted R&D support schemes, and second increased coordination of R&D activities. The latter has been achieved by the development of national strategies for R&D in oil and gas (first published in 2001 and updated every fifth year since). These strategies have been developed by broad involvement strategy formulation by industrial and research institutions. The ambition is to outline common goals and technological challenges facing the Norwegian petroleum industry, and based on this, to define prioritised technology and research areas. There have also been efforts to coordinate and streamline the innovation support offered by multiple public organisations (OG21, 2016). Research, development and innovation support is currently offered by four state organisations – the Norwegian Research Council, Innovation Norway (established 2003), Enova (established 2001) and Gassnova (established 2007) (see Chapter 3). The three latter organisations were all established after 2000, implying that the development of the institutional infrastructure for support of innovation is a fairly recent phenomenon.

The portfolio of R&D and innovation support programmes targeting upstream oil and gas have grown substantially during the last two decades, as have public allocations. Currently, there are more than 20 public schemes that fund R&D activities in the sector. The majority of these are not specifically petroleum-related programmes, but offer support to research, development and innovation activities in multiple industries. Several specific programmes targeted technological development in the petroleum industry. Figure 2.2 provides an overview of public allocation to specific petroleum-related R&D programmes (Petromaks, DEMO2000, Climit), as well as support to petroleum-related firms through general schemes such as SkatteFUNN and Innovation Norway.

As can be seen in Figure 2.2, public allocations have increased substantially since 2010, as has the industry's own R&D expenditures (Fjose *et al.*, 2014). As discussed above, this should be seen as interrelated with the new 'demand conditions' on a maturing shelf, which has led to a need for advanced technological solutions and substantial public and private R&D investments. The historically speaking high oil prices during recent years (see Chapter 1) is also relevant for understanding increased investments in R&D.

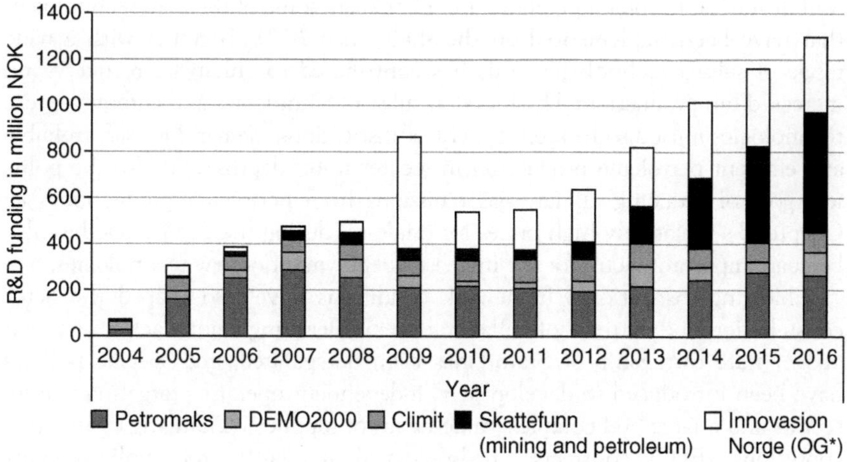

Figure 2.2 Allocations from petroleum-related R&D&I programmes by the Norwegian Research Council and Innovation Norway by year, in million NOK.

Sources: Norwegian Research Council (NRC), Project archive & Innovation Norway* 2005–2008 IN Data on industry 'Mining and petroleum extraction'. Since 2009, on sector 'Oil and gas'.

Co-evolution and transformation in the Norwegian petroleum innovation system

This chapter focuses on the coupled dynamics of technologies, actors, institutions and demand conditions in the Norwegian petroleum sector. The aim has been to shed light on the transformation of the petroleum innovation system during the last two decades. As has been discussed, the new demand conditions on a maturing petroleum region have to a large extent shaped the direction of the development of the sector. They have shaped policies and institutional developments and the actor constellations. They have also shaped the innovation efforts in terms of increased support and funding for innovation activities and the development of unique technologies that fit a mature sector where resources are located offshore in increasingly deeper waters. The geography of the Norwegian shelf itself and the nature of the natural conditions that contain the petroleum resources is therefore an important system element that needs to be taken into account. In this case, we clearly see a coupled dynamic between the maturation of production of petroleum resources with a focus on enhanced recovery and all other innovation systems elements that have supported this demand.

However, there are also other linkages between the system elements. The institutional infrastructure has shaped the development and selection of technologies in terms of the mode of development and deployment of technologies and the kinds of technology that have developed. 3D and 4D seismic, horizontal well drilling, multiphase pipelines and FPSOs are some of the core technologies that have been implemented on the shelf since 2000. Together with a wide range of subsea technologies, this has contributed to enhanced recovery and increased tail production. The focus on subsea technologies and carbon capture technologies must also be seen in light of institutional demand for safe, reliable and efficient petroleum production in greater water depths, and also the political goal of reducing climate gas emissions from petroleum production (see Chapter 3). Relatively high prices for crude oil during the last decade has also been an important factor for the increased deployment of new technologies.

Changing institutional framework conditions have also shaped the actor constellation and patterns of collaboration and learning among actors. As seen above, state ownership of Statoil has been deregulated, and specific policies have been introduced to develop new, independent operators targeting mature sectors and a large and complex value-chain of suppliers. Changing contractual conditions (EPCI contracts), technological standards and public support schemes to build competence, networks and 'technology readiness' of solutions held by Norwegian suppliers have all supported the development of the Norwegian supply industry. This process has also been guided by the long-term tradition of collaboration between operators, suppliers, research communities and policy bodies, which in this period has become even more fully institutionalised in a range of public-private arrangements for R&D and innovation support. These arguments are further elaborated in Chapter 3.

The increasing technology intensiveness of the sector has to be understood in relation to the development of actor constellations, particularly among the users (i.e. the operator companies). Institutional framework conditions have meant that to a large extent the system is dominated by one actor – Statoil – whose demands for new solutions steers the innovative efforts of the suppliers. Getting the support of the oil companies, and preferably a sign of interest and willingness to test and use the technologies, is *sine qua non* for the suppliers. Since Statoil has achieved such a dominant position, there could be less competition between new solutions in the future. However, allocation policies and petroleum tax policies have also supported increased diversification among operators where the intended goal has been to attract new oil companies to take care of tail production when the global oil giants exit the scene. Yet, these companies have less financial muscle and are less interested in support development and deployment of new technologies. Consequently, the need for public support for R&D has probably increased, and the petroleum tax regime whereby the state covers the majority of exploration costs has been instrumental.

To sum up, the last two decades have seen a transformation of the petroleum innovation system in Norway. This transformation is more along the lines of an evolution of the system rather than a marked departure from the past. As the sector has matured alongside the natural conditions on which the sector is built (i.e. the shelf), the innovation system has increasingly been characterised by a concentration of resources, but also increased diversity of actors – particularly among the suppliers. The supply industry has also grown due to the increased investment in advanced technological solutions, and particularly in areas such as subsea and reservoir knowledge. This has again been strongly supported by the development and institutionalisation of a public innovation support structure for oil and gas in Norway. This has also been important for suppliers' leap towards international petroleum markets (as is addressed in several chapters of this book) as well as for the diversification of the supply industry (see Chapters 11–13). This chapter demonstrates how demand conditions, the actor constellation, technologies and institutions are clearly linked in the case of the Norwegian petroleum sector and may thus serve as a general introduction to the more specific discussions in the following chapters.

Note

1 An overview of the different programmes: NORSOK is an abbreviation for 'Norsk Sokkels Konkurranseposisjon' or the competitive position of the Norwegian shelf. The NORSOK committee and NORSOK working groups consisted of representatives from the oil companies, main suppliers, engineering firms and the Government. The Norwegian Trade Union LO was also represented through NOPEF (Today, Industry and Energy). The NORSOK programme was initiated in 1993 and lasted until 2001. The first NORSOK standards were introduced in 1996 and have been refined, adjusted and developed until today. NORSOK's successor, KonKraft, also consists of the Norwegian Oil and Gas Association, the Federation of Norwegian Industries, the Norwegian Ship

Owners Association and the Norwegian Confederation of Trade Unions (LO). INTSOK – Norwegian Oil and Gas Partners – was established in 1997 by the Ministry of Oil and Energy, the Ministry of Industry and the Ministry of Foreign Affairs and in collaboration with the Norwegian Industry Association, The Ship Owners association, Norwegian Oil and Gas Association and Statoil, and Saga Petroleum og Norsk Hydro.

References

Acha, V.L. (2002). *Framing the Past and Future: The Development and Deployment of Technological Capabilities by the Oil Majors in the Upstream Petroleum Industry*. PhD thesis, University of Sussex.

Blomgren, A., Quale, C., Austnes-Underhaug, R., Harstad, A.M., Fjose, S., Wifstad, K., Melbye, C., Amble, A.B., Nyvold, C.E., Steffensen, T., Viggen, J.R., Iglebæk, F., Arnesen, T. and Hagen, S.E. (2015). *Industribyggerne 2015: En kartlegging av sysselsetting i norske petroleumsrelaterte virksomheter, med et særskilt fokus på leverandørbedriftenes eksportsysselsetning. [The builders of Industry 2015: A survey of employees in Norwegian petroleum related businesses, with a special emphasis on supplier industry employees related to exports.]* Report IRIS 2015/031. Stavanger: IRIS.

Dosi, G. (1982). Technological paradigms and technological trajectories: A suggested interpretation of the determinants and directions of technical change. *Research Policy*, 3, 147–162.

Engen, O.A. (2002). *Rhetoric and realities: the NORSOK programme and technical and organisational change in the Norwegian petroleum industrial complex*. PhD thesis, University of Bergen.

Engen, O.A. (2009). The development of the Norwegian Petroleum Innovation System: A historical overview. Mowery, D.C., Fagerberg, J. and Verspagen, B. (eds.), *Innovation, Path Dependency, and Policy: The Norwegian Case*. Cambridge: Cambridge University Press.

Fjose, S., Amble, I., Ramm, H.H. and Kroepelien, A.C. (2014). *Er tiden for de store teknologisprang over på norsk sokkel? [Is the time for great technology leaps over on the Norwegian continental shelf?] Report OG21*. Oslo: Menon Business Economics.

Malerba, F. (2002). Sectoral systems of innovation. *Research Policy*, 31(2), 247–264.

Malerba, F. (2005). Sectoral systems of innovation: a framework for linking innovation to the knowledge base, structure and dynamics of sectors. *Economics of Innovation and New Technology*, 14(1–2), 63–82.

Malerba, F. (2006). Innovation and the evolution of industries. *Journal of Evolutionary Economics*, 16(1–2), 3–23.

Nelson, R. and Winter, S. (1982). An *Evolutionary Theory of Economic Change*. Cambridge, MA: Harvard University Press.

Ryggvik, H. (2013). *Building a skilled national offshore industry. The Norwegian Experience*. Report NHO.

Rystad Energy (2015). *Internasjonal omsetning fra norske oljeserviceselskaper. [International sales from Norwegian oil service companies.]* Report for Ministry of Petroleum and Energy, 15. December 2015.

Vatne, E. (2013). *Den spesialiserte leverandørindustrien til petroleumsvirksomhet. Omfang og geografisk utbredelse i Norge. [The specialised supplier industry for petroleum activities. Scope and geographical distribution in Norway.]* SNF report 02/13.

Additional sources

Ministry of Oil and Energy [OED] (2015). Norsk verdiskaping i utbygging av petroleumsfelt [Norwegian value creation in the development of petroleum fields], report.

NOROG. (2016). NORSOK Analysis Project. Norwegian Oil and Gas Association, report.

Norwegian Petroleum. (2017). Number and diversity of companies. Retrieved 28.02.2018 www.norskpetroleum.no/en/developments-and-operations/number-and-diversity-of-companies/.

Norwegian Petroleum Directorate. (2016). Retrieved 28.02.2018 www.npd.no/no/Publikasjoner/Ressursrapporter/2016/Kapittel-5/

Norwegian Petroleum Directorate. (2017). Prekvalifisering [Pre-qualification]. Retrieved 28.02.2018 www.npd.no/no/Tema/Utvinningstillatelser/Temaartikler/Prekvalifisering/.

OG21. (2016). *Oil and gas for the 21st century*. Oslo: Norwegian Research Council.

White Paper 26 (1993–1994). St. meld 26 (1993–1994) Utfordringer og perspektiver for petroleumssektoren [Challenges and Perspectives for the Petroleum Industry on the Continental Shelf].

3 Innovation in the petroleum value chain and the role of supply companies

Erlend Osland Simensen and Taran Thune

Introduction

There is limited understanding of how innovation occurs in resource-based industries. Although there are several reasons for this, we believe that the issue has received too little attention within economics of development and more specifically within innovation studies. Within economics, there is a general idea that an abundance of natural resources has a negative impact on economic development (Mehlum, Moene and Torvik, 2006; Ross, 2012; Sachs and Warner, 2001). Researchers point to the 'curse' that resources impose on national economies, as countries with high share of their economy based on natural resources tend to have lower economic growth and less positive societal development. Moreover, within innovation studies, resource industries are seldom emphasised. Innovation research is frequently criticised for being too focused on manufacturing and hi-tech industries (Hirsch-Kreinsen, Jacobson, Laestadius and Smith, 2003; Robertson, Smith and von Tunzelmann, 2009; von Tunzelmann and Acha, 2005), and to emphasise 'boy's toys' rather than having a broader approach to innovation (Martin, 2016). The combination of scepticism towards natural resources and a bias towards manufacturing and high tech sectors has led to a situation with limited knowledge about innovation in natural resource industries (Andersen *et al.*, 2015; Ville and Wicken, 2012). This is problematic since resource industries in many countries are a major part of the economy. In the case of the natural resource industry we study in this book, the lack of knowledge is perhaps even more remarkable, as production of oil and gas is both a significant global industry and a highly technologically advanced process industry (Acha, 2002; Perrons, 2014).

In this chapter, we investigate innovation patterns and outcomes in the upstream oil and gas industry in Norway. To achieve this, we draw upon a perspective that argues that it is important to look at the linkages between resource-based sectors and the rest of the economy to understand development in resource-based industries. Rather than existing as detached islands, the resource-based industries are integrated and co-evolve with the broader economy (Andersen, 2012; Morris, Kaplinsky and Kaplan, 2012). In this line of work, the emphasis has been on the interaction between an 'enabling sector' and the resource sector (Andersen *et al.*, 2015; Ville and Wicken, 2012). For our analysis of the petroleum sector, the enabling sector would be companies that supply the industry with equipment, technology and services. A significant number of these companies are not classified as 'petroleum companies' in aggregated statistics because, according to standard industry classifications, they represent a variety of industries. What unites them is not product similarity, but rather the market they serve. Since this poses problems for analysts who study innovation by using industry-level data, most available studies of innovation in the petroleum sector have not been able to adequately capture the innovation inputs, processes and performance of the petroleum industry (Acha, 2002; Castellacci, 2008).

This book stresses that the Norwegian oil and gas industry is both technologically advanced and a highly innovative industry. This image contradicts an 'old' notion that resource-based industries in general, and the oil and gas industry specifically, is a 'low tech' industry. Due to their low share of R&D investment (measured as a proportion of revenues), resource-based industries, including industries such as mining and petroleum extraction, have been regarded as low- or medium-tech and therefore not especially innovative (von Tunzelman and Acha, 2005). This is, however, not a case of the Norwegian offshore petroleum extraction, we will argue, and maybe not for other resource industries either. There is a widely held perception among industry stakeholders that the oil and gas sector in Norway is responsible for the most advanced technological operations that have ever been developed and deployed in Norway (Blomgren *et al.*, 2015; Engen, 2009; Ryggvik, 2013). Due to specific environmental and regulatory conditions, collaboration between suppliers and oil companies to generate and exploit new and innovative technological solutions has been vital for the development of the Norwegian oil industry (see Chapter 2). This pattern of collaboration has led the Norwegian petroleum industry to become pioneers in locating, drilling and extracting petroleum in a relatively safe manner at great water depths and under extremely harsh weather conditions (Engen, 2009). This situation represents some kind of a puzzle. A sector that is regarded as a technological laggard in terms of investment in R&D is also seen as incredibly advanced when it comes to innovation. We address this puzzle by looking at the following research question: Does the petroleum sector have a unique way of organising innovation activities that distinguishes it from other industries?

To address this question, the chapter is structured as follows. In Part 2, we briefly discuss innovation in natural-resource-based industries, as a starting point. In Part 3, we look into the issue of innovation performance in petroleum economies and why it is difficult to measure. We exemplify this by looking at some key innovation parameters of the Norwegian economy. In Part IV, we turn to our dataset about innovation in the petroleum industry and the role of supply companies. In this section, we look particularly at three parameters for innovation: R&D and innovation investments, networks and innovation activities, and performance. We also compare the characteristics of the petroleum-related companies with selected other industries in Norway. Last, we summarise these findings and discuss what this means for the general knowledge about innovation in resource-based industries in general, and in oil and gas specifically.

Innovation in natural resource sectors

Natural resources are fundamental to the global economy. The largest share of energy production still relies on oil, gas and coal, and most vehicles run on carbon-based fuel. Moreover, naturally occurring materials are becoming increasingly important in fuelling a growing international economy and population with materials, technology and food. Most studies in the social sciences, however, have focused very little on these sectors' importance for the economy. Little is actually understood about how resource sectors have evolved, how innovation occurs in these sectors and how some countries have succeeded in using such industries for economic growth.

Many researchers have argued that an abundance of natural resources has negative consequences for national economies (Mehlum *et al.*, 2006; Ross, 2012; Sachs and Warner, 2001). The reason is that natural resources are thought to be a double-edged sword (Andersen *et al.*, 2015). On the one hand, income from selling commodities obviously brings revenues to the economy, but on the other hand, it can be detrimental for countries' economic development because of its ability to supress other sectors and also because increased labour costs weaken national competitiveness. The latter effect is often termed the 'Dutch disease'. The boom in the economy caused by the extra income from natural resources leads to inflation (currency appreciation), and the other products will thus be less price-competitive in an open economy. Since natural resources are believed to leave no long-term positive learning effect, the overall impact is detrimental for the long-term economic growth of a country. The resource is simply thought to be extracted until the source is depleted, leading to a fall in income and the country in recession.

The 'resource curse' concept has recently been criticised for being an over-simplification of a complex phenomenon, as it is based on a simple correlation between a country's abundance of resources and its economic development. Critics say that the causal effect may not be that straightforward, and that exceptions clearly exist. For instance, several countries have been successful in economic development based on natural resources, as is also seen in the recent history of Norway, Canada and Australia. Historical perspectives on economic growth argue that the abundance of natural resources has been important for most developed countries at some point in time (Ville and Wicken, 2012). Although many countries in recent years have struggled to develop despite an abundance of natural resources, this may thus not be a general trend. The concept can consequently falsely lead to a notion that resources themselves are bad for economic development, when it is the context that matters.

Recent research within innovation studies is also critical of the concept by pointing to the underlying assumption, i.e. that the manufacturing sector is superior to natural resource industries, as being wrong (Andersen *et al.*, 2015). Andersen *et al.* also criticise the assumption that natural resources do not need any linkages to the rest of the industry, but are there to be extracted (2015). They contend that science and technology has always been important for resource-based industries. Recent advances in science, such as in biotech, agriculture and materials, have been directly connected to the natural resource-based industries – making linkages to other parts of the economy even more important. They thus rather propose an explanation that the growth in extraction of natural resources is dependent on the deployment of advanced technologies.

The increased importance of knowledge is thus also valid for the extraction industries. To be able to access more (often increasingly inaccessible) resources, there is a need for sophisticated technologies and advanced knowledge. Hence, the industry has to work closely with other actors in technology development. This leads to a build-up of knowledge across industries, supported by a strong and high-earning sector that can afford to invest in research and development activities. The technological and organisational capabilities developed to support natural resource-based industries also have potential spill-over effects to other industries.

The effects of cross-industry knowledge spill-overs on innovation is, however, difficult to measure. Innovation is often measured on aggregated levels and then compared between countries. These aggregate numbers do not take into account that some countries have large resource-based sectors. Such countries often lag behind in international comparisons of innovation performance since traditional parameters of innovation emphasise innovation activities that occur in manufacturing industries. To exemplify this, we will turn to a case study of Norway. First, we will show that it scores relatively low on these indicators despite a high level of economic development, but that the reason for this is connected to the difficulties of measuring innovation in the resource-based industries.

Difficulties in measuring innovation in petroleum economies

Norway is one of the world's largest producers of petroleum resources and in recent decades has built up a significant supply industry (see Chapter 2). If we compare Norway to other countries, the oil and gas industry seems to have had a positive effect on manufacturing industry (see Chapter 1). Norway has increased its manufacturing sector more than comparable countries in the last 25 years; it is today 2.8 times larger than the 1990 level. To place this number in perspective, Sweden, Denmark and the Netherlands are all below 2.0, with the UK as low as 1.25 (Wicken, 2016). This seems to be associated with the rise of a significant petroleum supply industry, which also has significant export revenues. The capital goods industry is still the second most important export sector after oil ('Fuels'), much due to the emergence of a national petroleum supply industry.

However, some statistical parameters constantly show Norway as a 'laggard' on innovation; despite Norway's exceptional scores for GDP and productivity, the country scores lower than comparable countries such as Sweden, Denmark, Finland and Germany on innovation and R&D indicators (EU, 2016). This is not a new trend. For many years, Norway has had lower scores on these measures, and particularly on the level of R&D investments in industry.

The situation of high productivity and low innovation performance has been referred to as 'the Norwegian paradox' (Fagerberg *et al.*, 2009; Grønning *et al.*, 2008; OECD, 2007). There is a debate about the reasons for this, but many have claimed that this is due to the high presence of the oil and gas industry in the Norwegian economy. The resource- and commodity-based industries are usually regarded as 'low-' or 'medium-tech', which certainly also accounts for the oil and gas industry – an industry that has been found to have an R&D intensity which is lower than 1 per cent (von Tunzelmann and Acha, 2005). This means that the industry is considered as not having to deal with the development of advanced technological equipment or highly educated labour in order to operate. This explanation breaks with the aforementioned common perception of an increasingly technology-intensive and innovative oil and gas sector in Norway.

It is, however, more a case of measurement difficulties. Measurement of innovation is generally difficult, but even more so in complex, resource-based industries such as the petroleum industry (von Tunzelmann and Acha, 2005). The Norwegian oil and gas industry is no exception. Three characteristics of the Norwegian petroleum industry make innovation activities and performance in this sector particularly hard to measure.

First, for the oil companies, the large potential revenues lie in extracting the resources in an oilfield, and herein lies also their main focus. Since the petroleum industry is a process-oriented industry (Acha, 2002), what differentiates companies is not the product produced, but rather the processes of finding and extracting the resources. This affects commonly used innovation statistics such as the number of patents. Petroleum companies are not eager patentees and this tendency has been amplified in recent years as many companies have diversified and outsourced technology development to supply companies. Perrons (2014) found that 63 per cent of all innovations in the industry stemmed from the supply and service companies. As more of the risk and responsibility of oilfield developments now lies with the service and supply companies, the incentives for oil operators to patent have decreased. The oil companies increasingly procure technological solutions from a range of oil and gas supply companies. Consequently, it is beneficial for oil companies that the technologies are freely available since this will allow for increased competition among the supply companies, and hence lower the prices for the supplies and services they offer (Acha, 2002).

Second, and related to the first point, R&D activities are not customarily registered by the oil companies. Arguably, in large-scale engineering projects such as offshore drilling and operating oilfields, new technological solutions must be implemented with each project. This is due to the uniqueness of each project – each oil well is physically different from the next. These new technological solutions can often be new to the firm, industry or the world, and hence they are 'innovations' according to the standard definition of innovation (Mortensen and Bloch, 2005). The introduction of new technologies such as horizontal drilling and subsea systems for enhanced production exemplify that research and development is an important investment for oil and gas (O&G) companies. However, since all oil companies receive substantial tax refunds for exploration costs (see Chapter 2), oil companies have few incentives to report R&D costs. Large proportions of the research and development expenditure are simply reported as drilling and exploration costs. Furthermore, innovation in the Norwegian oil and gas industry is believed to rely more on industry collaboration and applied knowledge in everyday activities rather than purely scientific research and development (Isaksen and Karlsen, 2010). The innovation literature distinguishes between two ideal modes of innovation: Science, Technology and Innovation (STI) and Doing, Using and Interacting (DUI) (Jensen *et al.* 2007). STI is the innovation mode that is the easiest to measure, whereas DUI is more vague and not always captured by conventional measures of innovation. It seems that the innovation activities in the oil and gas sector may attribute more to the DUI than STI, an issue that we analyse in the next part of this chapter.

Third, general industry data has difficulties in singling out the full breadth of the O&G industry. In Norway, there are about 50 firms which are defined as 'operator companies', but activities are concentrated in a handful of these. Most statistics about the oil industry do not include the companies where large shares of the technology development occur – the supply and service companies (Perrons, 2014). The oil companies work together with a range of service and supply companies on the oilfields (Acha, 2002). These companies are often not classified as oil-specific companies, but as various forms of manufacturing and service firms. This is also valid for the many suppliers in the Norwegian oil industry. This is in line with what von Tunzelmann and Acha (2005, p. 411) state about the low- and medium-tech industries in general: 'the LMT industries resist easy classification, precisely because many of them are not very distinctive or singular in technological terms'.

To sum up, measurements of innovation activities in this sector are only interesting if we include the wider definition of the petroleum industry, i.e. include the petroleum supply companies. In addition, measuring traditional indicators of innovations such as R&D investment as a proportion of revenues or patenting does not capture the breadth of innovation activities in this sector nor the fact that most activities are widely distributed and are carried out in networks. Below, we turn to our analysis that has attempted to overcome these shortcomings.

Innovation activities and performance in the Norwegian petroleum sector

To understand how innovation occurs in this sector, we need micro-level data on innovation processes and performance of the broadly defined petroleum industry. To accomplish this, we have compiled a dataset that contains data on the innovation activities of the petroleum operator companies and a sample of petroleum supply companies that operate on the Norwegian continental shelf. We defined oil- and gas-related suppliers as companies that actively participate in any of the Norwegian O&G industry associations and/or industry clusters, and by this define themselves as an oil- and gas-related company. Our sample consists of 620 companies. In addition to collecting basic firm level data from public databases and public accounts data, we collected information about investments in R&D and innovation, innovation models and results. We did this by combining our firm level database with data from two other datasets: data from the firm level Community Innovation Survey (CIS) and data from participation in research and innovation projects maintained by the Norwegian Research Council. The latter data is used to analyse the topology of the innovation networks in petroleum and the role of supply companies within such networks.

R&D investments in the Norwegian petroleum sector

Public-supported R&D in upstream petroleum represents only a small share of the overall investment. According to calculations made by Fjose *et al.* (2014), public investment represents about 10 per cent of the total investment in R&D and innovation. Moreover, their estimates indicate that private R&D investments in the petroleum sector are split fairly equally between the oil companies and the petroleum supply industry. Their data show that R&D investments grew substantially after the millennium in both segments. In particular, the growth in investments in R&D was substantial in the period 2010–2013. Prior to this (in the early 2000s), growth was more modest, but in this period the R&D investments by the oil companies grew by 44 per cent and 50 per cent in the supply industry (Fjose *et al.*, 2014). After 2013, R&D investments by oil companies have dropped, and in all likelihood by a great deal in the petroleum supply industry. Investments made by public sources have grown substantially since 2014, and particularly in 2016 and 2017, to remedy the fall in private investments due to the significant drop in the oil price globally (see Chapter 2).

These data indicate that both public and private investments in research and development activities have increased significantly over time. Moreover, the volume of investment indicates that industries which make up the petroleum industry complex in Norway are not laggards in R&D investment. To make this point even more clear, we will look at the R&D investments in petroleum compared to other Norwegian industries. Figure 3.1 compares R&D investments in two dimensions – the number of employees working in R&D (number of man-years of employees with PhD qualifications), and overall annual

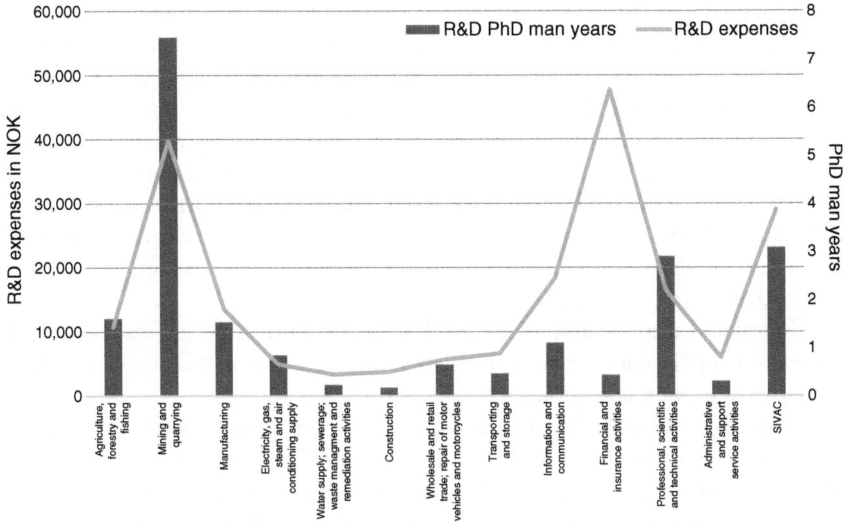

Figure 3.1 Average R&D expenses in Mio NOK and PhD 'man years' across Norwegian industries.

Source: CIS, 2014.

investment in R&D activities. Here, the main sectors in Norway are compared, and in addition, we have added data for 'petroleum supply companies' based on our database of designated petroleum suppliers (called 'SIVAC' in the figure after the name of the database).

Quarrying and mining – the industrial category where oil companies are located – has the most man-years engaged in R&D followed by scientific and technical service providers (applied research institutes and consultancy and engineering companies) and petroleum suppliers. Taken together, petroleum supply companies and petroleum companies employ the majority of PhD trained R&D workers in Norwegian industry.

Total R&D expenses are also among the highest for Norwegian industries. Taken together as the 'petroleum related sectors' (both petroleum companies and petroleum supply companies) is that industrial sector which spends the most on R&D of all Norwegian industries. These data were however collected in 2014 (the last available data) and investments in R&D activities and employees have probably decreased somewhat since. Nonetheless, the overall image is that the petroleum sector (including suppliers) is among the most R&D intensive industrial sector in Norway.

Innovation activities and performance in the petroleum sector

Although R&D expenditure is frequently used as an indicator for innovation, we assume that many innovation activities in the petroleum industry are not defined as R&D. Rather, the industry develops and deploys new technologies and new processes through exploratory activities and in close collaboration with suppliers and users of technologies (Acha and Cusmano, 2005; Perrons, 2014).

In order to capture a broader spectrum of innovation activity, we draw upon data collected through the CIS. This is a survey which investigates innovation in Norwegian firms biannually. We measured innovation as total turnover from new or improved goods and services, where both new to market and new to the company were included. Furthermore, several of the items in the survey allow us to look at different forms of innovation where product and service innovation is a measure of whether the company had carried out product or service innovation between 2010 and 2012.

The overall results show that for all forms of innovation measured in CIS, oil-related supply companies score higher than average in Norway, as well as higher than other comparable sectors (see Table 3.1.)

As seen in the table, petroleum supply firms report high above average for Norwegian companies for all types of innovation. This even counts for product innovation, the only measure where manufacturing firms score higher than the petroleum suppliers. On service innovation, however, oil-related suppliers score significantly higher than manufacturing firms, and also above the average for all Norwegian firms.

We have also looked at two more variables representing two typical modes of innovation: science, technology and innovation mode of innovation (STI), and the DUI mode of innovation (Jensen *et al.*, 2007). The first is linked to investments in technology and highly skilled employees. The latter, the DUI mode, is a softer type of innovation that many innovation researchers believe to be as important as the STI mode. The DUI mode depends on experience-based knowledge, often through more informal processes of learning and experimentation, frequently involving users and producers of technologies (i.e. Doing, Using, Interacting). The STI indicator refers to R&D investments between 2010 and 2012 The DUI mode is more complex. Here we have added six items from the CIS. There were four indicators related to collaboration with customers, suppliers and competitors, and also within the company. Two more indicators of the DUI mode were investments in technology development (non-R&D) and training of technology-related personnel. What is interesting is that the O&G-related suppliers score high on both the STI and DUI modes of innovation. This is an indication that the petroleum sector leans on both formal innovation inputs as well as collaboration and informal learning processes to innovate.

In addition to simply comparing the means of these indicators, we tested the relative importance of these modes for innovation performance by sector. By performing regression analyses, we found that the DUI mode was more important for innovation performance in the petroleum supply companies compared to other sectors. This result was valid for a broad definition of innovation (income from both radical and incremental innovation). We also tested whether the DUI mode increased in importance in service and product innovation. We found support for an extra effect on service innovation, but we did not

Table 3.1 Average of innovation indicators for oil-related suppliers companies vs. other sectors. The table compares different measures of innovation across the petroleum supplier industry and sectors in Norway

Sector	N	*Innovation*	*Product innovation*	*Service innovation*	*Doing, using and interacting*	*Science and technology indicators*
Petroleum suppliers	275	5.402 (0.460)	0.273 (0.0269)	0.135 (0.0206)	2.689 (0.166)	1.029 (0.0237)
All companies	3,192	3.773 (0.114)	0.195 (0.00702)	0.106 (0.00545)	1.745 (0.0399)	0.730 (0.0237)
Manufacturing	1,155	4.971 (0.209)	0.329 (0.0138)	0.0398 (0.00576)	2.181 (0.0702)	0.939 (0.0453)
Financial and insurance activities	160	1.550 (0.372)	0.025 (0.0124)	0.144 (0.0278)	0.687 (0.123)	0.444 (0.0886)
Professional, scientific and technical activities	324	4.581 (0.374)	0.201 (0.0223)	0.182 (0.0215)	2.525 (0.139)	0.861 (0.0755)
Constructing	231	0.402 (0.162)	0.0130 (0.00747)	0.0216 (0.00960)	0.402 (0.0767)	0.186 (0.0432)
Mining and quarrying	159	2.760 (0.477)	0.113 (0.0252)	0.0818 (0.0218)	1.578 (0.190)	0.767 (0.112)

Source: CIS 2014 and own database.

find this for product innovation performance. These results imply that this sector is focused more on services, and furthermore, that collaboration and investment in technology and experienced-based knowledge is particularly important for innovation performance in services.

The importance of collaboration in this industry cannot be stressed enough, where technology development and deployment in the industry is organised as an 'ecosystem' (Acha, 2002). Major oil companies procure large, complex projects from large service and supply firms. These companies are system operators which purchase technologies and services from a range of sub-suppliers and technology vendors. Each alteration in a process or component when so many actors are involved requires a well-functioning collaborative system. For instance, if an under-supplier suggests a change in one of the components in an offshore installation, this can potentially affect other parts of the system. Collaboration is thus crucial for such a system to work and the industry has several mechanisms forums where technology development and implementation occur in collaboration between several stakeholders.

Innovation networks and collaboration across industries

The oil and gas companies organise their technology development mainly in three different settings. One is through the joint industry programmes usually initiated by the companies themselves. Here, companies and researchers meet to create projects designed to solve common problems. Second, O&G companies develop technological solutions and new processes on the oil and gas fields, mainly by implementing new technologies in an offshore installation or through changing the combination of existing technologies. Last, there are the governmental funded research programmes in Norway organised and channelled through the Norwegian Research Council (see Chapter 2).

In the following, we briefly describe the petroleum research network in Norway and how it has evolved during the last decade. To describe the network composition and development, we acquired data on government supported research projects from the Norwegian Research Council. Here, we identified the O&G-related research programmes and based on these research projects, we constructed a social network of the participating actors.[1] Each actor was a company, a research institution or a government institution.

This network obviously does not represent all interactions between actors in the innovation system surrounding the Norwegian upstream O&G industry. For instance, the database does not cover the research initiatives that were declined by the Research Council, neither does it cover private R&D initiatives such as joint industry projects funded by the industry. However, many projects have been jointly funded by the government, and do appear in these networks.

This network analysis claims thus not to analyse the full depth of industry interaction, but it would be reasonable to assume that it reflects the strongest actors in the Norwegian petroleum-related innovation system. What we do know is that the actors analysed here have had frequent R&D-related contact during the period and that this network is the most complete network of petroleum actors in a country hitherto. We divided the network into three different time periods in order to analyse the evolution of the network and to be certain that the actors do interact in network (See Figure 3.2).

The most telling result from these network graphs is that many supply companies are involved in research projects, even though none of these are the most central nodes. If we look at the core of the network, three highly central actors dominate the network in all three periods – these are two technological research institutions, and a dominating oil company. We can also observe strong ties between these core actors, meaning that they frequently work together. Hence, they are the actors which participate in most funded research projects, and their neighbours are very centrally positioned. The two research institutions gain centrality throughout the period, whereas the oil company participates less in research projects as time evolves. In this representation of the network, the anticipation lies with the supply companies, so we have toned down the dominance of the other actors by colouring them white. The overall structure is steadily centralised, but seems to open more up as time progresses. The supply companies, represented as black nodes in the figure, seem also to be more co-located in the latter two periods; that is, in the first period, the supply companies are more scattered throughout the graph. This indicates that in the latter two periods the projects involve more collaboration between supply companies.

One reason for this development is that there seems to be an increasing division in types of project. One is where industry-relevant research projects cluster together around the supply companies; this could be a project that focuses on implementing new technology on the Norwegian shelf. The other type of project is more closely related to emerging research topics such as how to deal with climate gas emissions from oilfields which are oil-related, but funded without many supply companies participating.

We undertook a content analysis of central words and terms in this industry, which confirmed this assumption. The first period involved only one project about Carbon, Capture and Storage (CCS), whereas they were numerous in 2009–2012 (23) and in 2013–2015 (21), due to an increased focus on climate change from around 2006–2007, which also resulted in the infamous 'moon landing' project announced by then prime minister Jens Stoltenberg. Stoltenberg announced that Norway should implement and commercialise CCS technology ready for the new gas power plant at Mongstad, which at the time was under construction. This led to increased funding for similar projects in the following years. These types of projects were more likely to exclude Norwegian companies as research partners, and were driven by research institutions, not private companies.

As mentioned above, the Norwegian supply companies are numerous but not highly central in the network, neither when measured as average scores nor as individual company scores. Hence, each supply company participates on fewer occasions in research projects. This is somewhat expected since they represent smaller organisations than research institutions and oil companies, and their

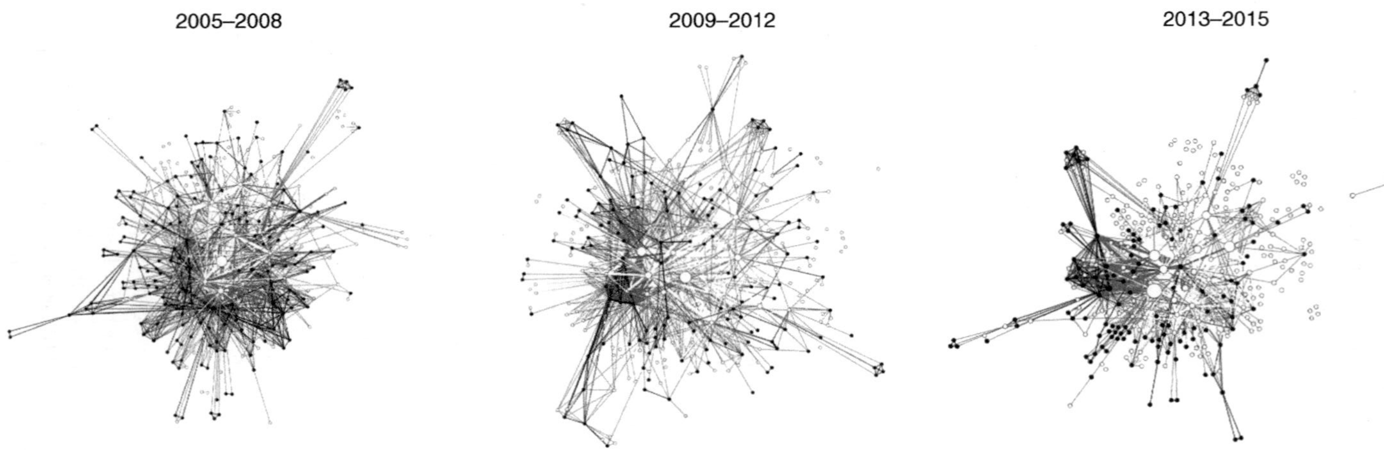

Figure 3.2 Three timespans of research network of actors. Black nodes are Norwegian registered companies, whereas white nodes represent all other types of actor.

Source: own database.

technological domain is narrower than many of the other actors in the network. Although their positions seem peripheral in the network, they possess important roles. In fact, most research projects include a supply company which also tends to be the project leader (Figures 3.3 and 3.4).

In other words, the funded projects are mainly classified as one of two types: those managed by supply companies that deliver technologies to the petroleum projects, or those managed by research institutions in collaboration with the petroleum-related companies. Supply companies are often responsible for developing their own technologies. They often include research institutes and the customers, i.e. the oil companies, in the projects. It is clear that the oil companies' roles, despite their high participation rate in the network, are sponsors and contributors rather than technology owners. Concerning the general composition of the actors, the companies dominate in numbers in all three periods given above and the composition of the actors is stable. One exception is that there seems to be a large increase in participation by foreign research institutes – an increase of almost 300 per cent from 2005–2008 to 2013–2015. Hence, petroleum-related research has had an increase in internationalisation over the three timespans. This tendency is probably driven by a general increase in internationalisation, but also the increased interest by Norwegian firms in foreign offshore petroleum markets as well as foreign offshore markets' increased interest in Norwegian competence and experience.

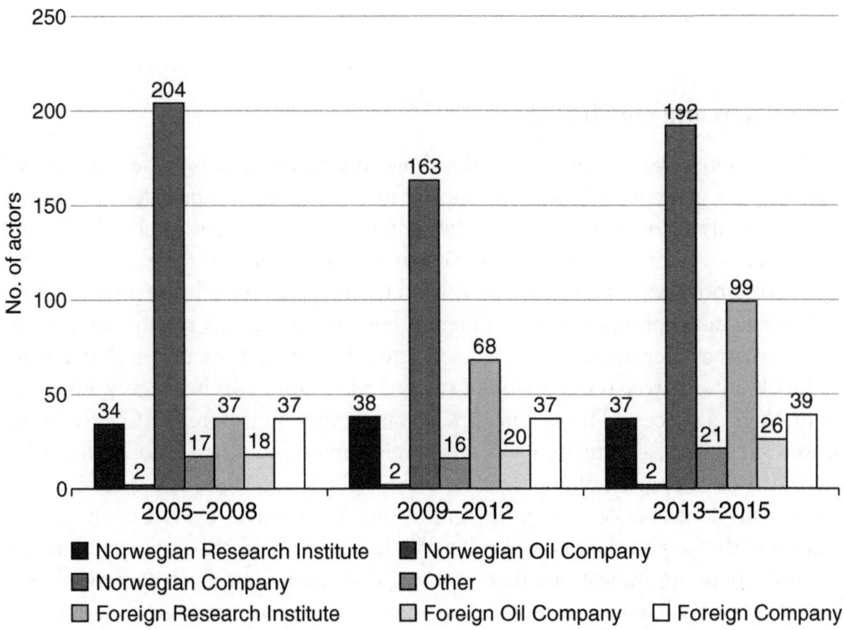

Figure 3.3 Composition of type of actors in the network in three different periods.

Source: own database.

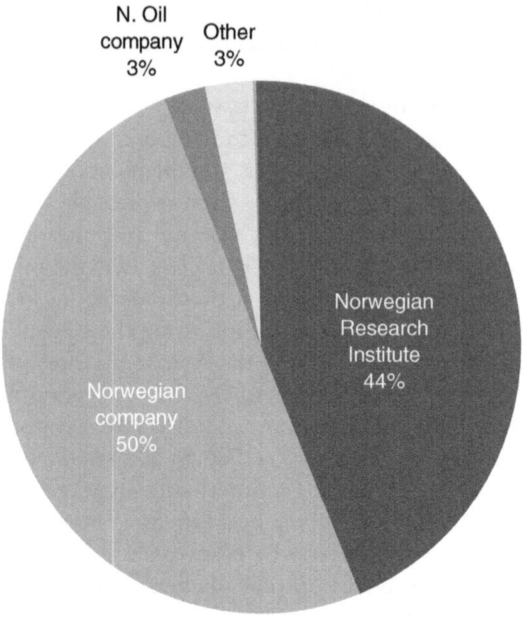

Figure 3.4 Type of institutions that were project leaders of the research projects.

Source: own database.

Discussion and conclusion

In this chapter, we set out to describe how innovation occurs in resource-based industries in general and more specifically in the Norwegian petroleum industry. Several analysts have pointed to the unique way that new technologies are developed and deployed in the petroleum sector, particularly the collaborative nature of innovation in upstream petroleum where large and affluent oil companies and a wide range of suppliers fulfil different but complementary roles in an innovation ecosystem. Recent literature has suggested that natural resource industries can be highly innovative, as extraction of resources in many cases is a highly advanced technological process. The technological complexity of offshore O&G extraction, as well as the increased tendency of outsourcing among the large O&G companies, drives the development of an 'enabling sector' (Andersen *et al.*, 2015). This sector is the main developer of a range of technologies and forms of expertise – in collaboration with the petroleum companies. We have drawn on this perspective to see whether there are indications that innovation in petroleum is characterised by a high degree of collaboration between operators and suppliers of advanced solutions, and the importance of this collaboration for innovation output.

To look at this empirically in the case of the Norwegian petroleum sector, we have developed a dataset that encompasses both oil companies and their suppliers, and we looked at both traditional indicators such as R&D statistics and data sources which capture advanced innovations which are not R&D-based. Our analysis indicates that both traditional STI-indicators (manpower and investments in R&D) and DUI-indicators confirm that the Norwegian petroleum industry and its supply industry score well above average. The supply industry seems to be particularly good at advanced service innovations, and has a higher than average DUI pattern of innovation, as well as a larger effect from this innovation mode. The collaborative nature of innovation is also visible in the high degree of networking where supply companies are involved in the majority of publicly funded R&D projects within petroleum. Further information about the importance of networking also at regional levels is found in Chapter 4.

Overall, it seems fair to say that the data indicates that the petroleum sector, also including the supply sector, is highly innovative. Moreover, contrary to some other recent studies, we do not find that the supply companies have taken over the technological leadership (Perrons, 2014). One reason for this is these studies focus on a sample of large oil and supply companies that operate globally. Our study focuses on another stratum of the petroleum industry. Rather than focusing on a sample of actors, it focuses on all the actors within one petroleum region, and also includes technology development and deployment data that is not R&D.

We did expect to see the shift in innovation capability from operators to suppliers, but our data does not support this. The results of our analyses (as well as most chapters of this book) are that rather than a marked shift, we see a dispersion of technological capabilities in an increasingly wide network of actors. The operators – although they do not develop technologies themselves, are still involved in 'everything' and have matching capabilities across most areas of technology. This is probably due to the technological complexities of petroleum operations, the need for system integration and the strong focus on risk, control and liability that the operators of fields will have.

These results also indicate that most of the O&G industry has attracted the many technologically advanced companies and personnel in Norway. There are signs that the comparably low innovation scores of Norway are due to the increased emphasis on collaborative problem-solving rather than investment in typical technological parameters (patenting and R&D). However, it is difficult to prove this directly. What seems to be more certain is in the discussion about the effect of natural resources on national economies. Here, Norway emerges as a convincing example of a country that has been able to build a strong economy through creating spill-overs from natural resource industries. Moreover, the spill-over has led to the building of a strong national supply industry with increasing export revenues (see Chapters 6 and 7).

Taken together, the findings in this chapter have some important implications for the understanding of economic development of resource-rich countries. If it is so that natural resources per se are not to blame for low innovation capability and poor economic development, policies should support the build-up of advanced capabilities alongside resource sectors rather than conclude that countries need to achieve a 'higher' level of manufacturing/industrial activity in order to become more developed. This perspective is also important for how we view natural resources' role in economic growth, and motivates us to perform further empirical studies on the topic.

Even though Norway has been very successful in recent years, it does not mean that the same model will be the correct model for the future. A maturing continental shelf and increased competition from renewable energy makes the long-term profitability of the O&G sector uncertain. New policies for diversification of the economy partly based on existing capabilities are thus important for Norway in the years to come (see Chapters 11 and 12). It is important to note that the starting position is good; Norway has the financial power to invest in new industries and has, as we have seen, world-leading capabilities in organising complex and technology-intensive projects.

Note

1 The research programmes are PETROMAKS 1, PETROMAKS 2, PETROSAM, Gassmaks, DEMO2000, OG, CLIMIT, CO2-HAND.

References

Acha, V. L. (2002). *Framing the past and future: the development and deployment of technological capabilities by the oil majors in the upstream petroleum industry*. PhD thesis, University of Sussex.

Acha, V., and Cusmano, L. (2005). Governance and co-ordination of distributed innovation processes: patterns of R&D co-operation in the upstream petroleum industry. *Economics of Innovation and New Technology*, 14(1–2), 1–21.

Andersen, A. D. (2012). Towards a new approach to natural resources and development: the role of learning, innovation and linkage dynamics. *International Journal of Technological Learning, Innovation and Development*, 5, 291–324.

Andersen, A. D., Johnson, B. H., Marín, A., Kaplan, D., Stubrin, L., Lundvall, B.-Å. and Kaplinsky, R. (2015). *Natural resources, innovation and development*. Aalborg, DK: Universitetsforlag.

Blomgren, A., Quale, C., Austnes-Underhaug, R., Harstad, A. M., Fjose, S., Wifstad, K., Melbye, C., Amble, A. B., Nyvold, C. E., Steffensen, T., Viggen, J. R., Iglebæk, F., Arnesen, T. and Hagen, S. E. (2015). *Industribyggerne 2015: En kartlegging av sysselsetting i norske petroleumsrelaterte virksomheter, med et særskilt fokus på leverandørbedriftenes eksportsysselsetning. [The builders of Industry 2015: A survey of employees in Norwegian petroleum related businesses, with a special emphasis on supplier industry employees related to exports.]* Report IRIS 2015/031. Stavanger: IRIS.

Castellacci, F. (2008). Innovation in Norway in a European perspective. *Nordic Journal of Political Economy*, 34, 43.

Engen, O. A. (2009). The development of the Norwegian petroleum innovation system: a historical overview. J. Fagerberg, D. C. Mowery, and B. Verspagen (eds.), *Innovation, Path Dependecy and Policy. The Norwegian Case*, 387. Oxford: Oxford University Press.

Fagerberg, J., Mowery, D. and Verspagen, B. (2009). *Innovation, Path Dependency, and Policy: The Norwegian Case*. Cary: Oxford University Press.

Fjose, S., Amble, I., Ramm, H. H. and Kroepelien, A. C. (2014). *Er tiden for de store teknologisprang over på norsk sokkel?* [Is the time for great technology leaps over on the Norwegian continental shelf?] Oslo: Menon Business Economics.

Grønning, T., Moen, S. E. and Olsen, D. S. (2008). Low innovation intensity, high growth and specialised trajectories: Norway. *Small Country Innovation Systems. Globalization, Change and Policy in Asia and Europe*, 281–318.

Hirsch-Kreinsen, H., Jacobson, D., Laestadius, S. and Smith, K. (2003). *Low-tech industries and the knowledge economy: state of the art and research challenges*: SINTEF STEP Group.

Isaksen, A., and Karlsen, J. (2010). Different modes of innovation and the challenge of connecting universities and industry: case studies of two regional industries in Norway. *European Planning Studies*, 18(12), 1993–2008.

Jensen, M. B., Johnson, B., Lorenz, E. and Lundvall, B. Å. (2007). Forms of knowledge and modes of innovation. *Research Policy*, 36(5), 680–693.

Martin, B. R. (2016). Twenty challenges for innovation studies. *Science and Public Policy*, 43(3), 432–450.

Mehlum, H., Moene, K. and Torvik, R. (2006). Institutions and the resource curse. *The economic journal*, 116(508), 1–20.

Morris, M., Kaplinsky, R. and Kaplan, D. (2012). 'One thing leads to another' – Commodities, linkages and industrial development. *Resources Policy*, 37, 408–416.

Mortensen, P. S. and Bloch, C. W. (2005). *Oslo Manual-Guidelines for Collecting and Interpreting Innovation Data: Proposed Guidelines for Collecting and Interpreting Innovation Data*: Organisation for Economic Cooporation and Development, OECD.

Perrons, R. K. (2014). How innovation and R&D happen in the upstream oil and gas industry: Insights from a global survey. *Journal of Petroleum Science and Engineering*, 124, 301–312.

Robertson, P., Smith, K. and von Tunzelmann, N. (2009). Innovation in low- and medium-technology industries. *Research Policy*, 38(3), 441–446.

Ross, M. (2012). *The oil curse. How Petroleum Wealth Shapes the Development of Nations*. Princeton, NJ: Princeton University Press.

Ryggvik, H. (2013). *Building a skilled national offshore oil industry: The Norwegian experience*. Oslo: The Confederation of Norwegian Enterprise (NHO).

Sachs, J. D. and Warner, A. M. (2001). The curse of natural resources. *European economic review*, 45(4), 827–838.

Ville, S. and Wicken, O. (2012). The dynamics of resource-based economic development: evidence from Australia and Norway. *Industrial and Corporate Change*, 22(5), 1341–1371.

Von Tunzelmann, N. and Acha, V. (2005). Innovation in low-tech industries. J. Fagerberg, D. C. Mowery and R. R. Nelson (eds.), *The Oxford Handbook of Innovation*, 407–432. Oxford: Oxford University Press.

Wicken, O. (2016). 12 Industrial diversification processes and strategies in an oil economy. *Economic Diversification Policies in Natural Resource Rich Economies*, 295.

Additional sources

EU. (2016). European Innovation Scoreboard. Retrieved 28.02.2018 http://ec.europa.eu/growth/industry/innovation/facts-figures/scoreboards_en.

OECD. (2007). *OECD Economic Surveys: Norway – Volume 2007 Issue 2*. Paris: Organisation for Economic Co-operation and Development.

4 Knowledge networks and innovation among subsea firms

Nina Hjertvikrem and Rune Dahl Fitjar

Introduction

Oil and gas on the Norwegian Continental Shelf involves deep waters, and the use of traditional platforms and divers is challenging because of depths and the rough seas. Equipment installed on the seabed therefore improves the opportunities to search for and produce oil and gas safely. The natural conditions of oil production in the North Sea has presented companies with enormous technological challenges, as described in Chapter 2. One important challenge pertains to the installation, maintenance and operation of equipment on the seabed. In the early stages of the Norwegian offshore industry, divers conducted most such operations. However, this was associated with safety issues and both short-term and long-term health risks. Many resources were therefore devoted to the development of subsea technology commencing in the 1980s.

In the 1990s, breakthroughs occurred in subsea technology making it possible to extract oil more safely and in deeper seas. Partly as a result of this, the number of fatal work accidents declined. Most subsea accidents between 1980 and 1990 involved divers (Ryggvik, 2017), but relatively few during the last ten years (Petroleumstilsynet, 2016). By reducing the need for divers, the automation of subsea production made offshore oil production safer. One of the big steps forward for the subsea sector was an innovation, 'Clamp Connector', which made it possible to connect installations on the seabed without the use of divers. This innovation has been taken up globally (Aker Solutions, 2017). Some of the firms in the subsea sector started as providers of divers. Today, remotely operated vehicles (ROVs) and other equipment that can be operated from shore or from ships have largely replaced the need for divers. Due to these developments, Norwegian suppliers are today seen as world-leaders in subsea technologies.

In this chapter, we examine innovation processes and knowledge networks in the subsea industry in Norway's oil capital, Stavanger, and its surrounding region, Rogaland. Being located in a region with a high density of competitors, suppliers and other interrelated industries is said to be conducive to 'localized knowledge spillovers'. These may materialise from collaboration between firms (Bathelt and Turi, 2011; Storper and Venables, 2004), labour mobility

(Timmermans and Boschma, 2013) and the opportunity to constantly monitor and compare with other firms (Bathelt *et al.*, 2004). The subsea sector in Rogaland is located in a region with many competitors and many interrelated industries. Subsea firms have the opportunity to collaborate with, recruit from and monitor other firms in their industry, as well as firms in related industries.

In 2015, we interviewed 30 of 31 firms operating in the subsea industry in Rogaland – the region in which Stavanger is located – collecting data on their innovation output and innovation processes. This included collecting full network data on their collaboration networks.

We find that subsea is a highly innovative industry. In total, 83 per cent of firms report product innovations in the last three years, and 63 per cent report new-to-market innovations. Innovation is mainly a result of problem-solving in response to customer needs, and is heavily engineering-based. Subsea firms typically make tailor-made products for their customers, and compete on performance and problem-solving ability more so than cost. This makes collaboration with oil operators paramount, and all subsea companies have multiple connections with oil operators. However, the collaboration network within the subsea industry is also very dense. An average subsea firm collaborates with seven other firms in the local subsea industry. Projects often involve various subsea firms, and firms frequently supply components for each other's products. This creates fertile conditions for knowledge exchange in the industry, but also carries a risk of lock-in.

Norwegian subsea industry clusters

The subsea industry supplies products and services for use between the seabed and the surface in offshore oil and gas production. The industry includes firms which produce subsea equipment, firms which install equipment and firms which maintain existing subsea equipment. As such, the industry includes firms engaged in various different types of activity and which supply goods and/or services to other subsea firms, to general oil service companies or directly to the oil operator companies.

Norway has a large share of the subsea market, accounting for around half of the global market. The firms can be divided into two types, one specialising in technology and development, and the other specialising in planning and installation offshore (Reve and Sasson, 2012). The firms in the Norwegian subsea sector are mainly located in three regions: Buskerud, Hordaland and Rogaland. Hordaland and Rogaland are both located on the west coast, close to the offshore activity on the Norwegian Continental Shelf. The main cluster of subsea firms in Norway is found in the axis from Kongsberg (in Buskerud county) to Oslo, known as Subsea Valley. This includes four of the five largest Norwegian subsea companies. This cluster mainly specialises in development of subsea technology. Another important cluster is in Hordaland county, world-leading in operating, maintaining and modifying subsea equipment. This cluster was awarded Global Centre of Expertise status (GCE Subsea) by the Norwegian cluster programme in 2015. The cluster organisation GCE Subsea consists of more than 100 companies and organisations. The subsea industry in Rogaland is also substantial, accounting for around 10 per cent of oil service employment in the most oil-intensive region of Norway (Blomgren *et al.*, 2015; see Chapter 10). While Buskerud and Hordaland export more subsea equipment, the subsea firms in Rogaland (especially compared to Buskerud) are more hands-on and involved in practical planning and installation of subsea operations. The Stavanger region houses the biggest petroleum cluster in Norway, including all parts of the oil industry value chain with its various oil service and supply sectors (see Chapter 10).

The subsea industry in Rogaland

This chapter is based on a population study of all firms specialising in subsea technology in Rogaland, which was conducted in early 2015. In order to identify subsea firms, we used a population database of all oil-related firms in Norway compiled by the International Research Institute of Stavanger (IRIS) (Blomgren *et al.*, 2015). The database included all unique subsea firms registered in Rogaland county with five employees or more. This resulted in a population of 31 firms active in the region's subsea industry. These were all contacted for interviews, and we conducted interviews in 30 firms, a response rate of 96.8 per cent. Data collection was based on personal interviews lasting for around 45 minutes in each case. The interviews were structured around a questionnaire with a combination of closed- and open-ended questions, including network data on the firm's collaboration partners, recruitment and inspiration sources. In half of the cases, we interviewed the CEO of the firm. The other respondents comprised regional branch managers, technical managers or others in managerial positions. Data collection was undertaken shortly before the fall in oil prices which caused economic turmoil in the Norwegian oil industry, including the subsea industry in Rogaland. The years prior to the study were characterised by high oil prices and an unprecedented period of growth in the industry. As such, the data represents the innovation activities and processes at the height of the oil boom in the Stavanger region.

Table 4.1 shows characteristics of the interviewed firms. There is wide variation in the size, age and ownership structure of the firms, ranging from small local workshops to large multinational enterprises. Smaller firms frequently offer a specific high-technology product and typically employ a high proportion of engineers. Some of the subsea firms are fairly old, predating the era of oil and gas exploitation in Norway. One firm started as a small family business renovating cars, another was a local workshop, and a third used to make and repair agricultural equipment. Some of the firms offered diving services and seized the opportunity to move into a new market in the oil and gas industry, changing from divers to ROVs and related services in the process. Some firms mention their experience from diving as a strength even though they have now replaced divers with ROVs. Two multinational firms in Stavanger still offer diving services.

Table 4.1 Characteristics of the firms in the subsea sector in Rogaland (based on interviews)

Age		*Size*		*Head quarter*		*Share employees with higher education*		*Share of employees in R&D*	
Years	N		N	*Location*	N	%	N	%	N
<5	5	Small	8	Regional	13	>90:	6	>30	1
5–10	9	Medium	15	Norway:	6	70–90:	0	>20	2
10–20	4	Large	7	Europe:	8	50–70:	7	>10	2
20–30	5			USA:	3	30–50:	3	>5	3
>30	7					<30:	14	>1	13
								0	9

Innovation in the Rogaland subsea industry

A large share of firms in the subsea industry in Rogaland report innovation in terms of new products or processes as shown in Table 4.2. In total, 25 of 30 firms (83 per cent) report having introduced new products or services during the three years preceding the survey. Nineteen of these also report new-to-market innovations which we use as an indicator for radical innovation.

Innovation in the subsea industry is typically based on customers' needs and problem-solving where the customer (typically an oil operator) presents problems for which subsea firms develop solutions. Innovation is mainly incremental, typically in the form of small scale tailor-made products at a high cost. Some problems might force firms to find completely new solutions by using new technology or new materials in the product, resulting in a product or services new to the market. Product development is based on close contact between suppliers and customers throughout the development phase. Several subsea firms stated that their main reason for being located in Rogaland was the need for communication with customers, sometimes 24/7. Others maintained that their customers (operators) demanded their presence in Stavanger because of a need for daily face-to face dialogue and discussion on how to solve problems and to be able to react at short notice.

Innovation processes in the subsea sector have a strong element of learning-by-doing. As product development mainly takes the form of problem-solving, it is often not reported as R&D activity in tax returns. Consequently, R&D activity in this sector is probably higher than reported in official statistics (see also Chapter 3). The industry has also taken advantage of competence from other sectors in the oil and gas industry with deeper knowledge on specific challenges (like pressure and temperature), through recruitment from other sectors like seismic services, drilling, and measurement (Reve and Sasson, 2012).

Among the interviewed firms, 23 have employees devoted to product development. However, the proportion of employees working specifically on product development varies greatly. In one of the larger firms, just one employee was engaged in product development while in some of the smaller firms a majority of employees were engaged in this activity. Most firms state that they do not have

Table 4.2 Results for the different types of innovation activity in the subsea firms (based on interviews)

Results of innovation activity N = 30	*YES*	*NO*
Product innovation	25	5
New-to-market product innovation	19	11
Discoveries	23	7
New processes	19	11
New business strategy	26	4
New business structure	25	5
New marketing strategy	15	15

the resources to enable their employees to do research only. This is too costly, and R&D is mostly undertaken as part of problem-solving for a customer. The case study in Chapter 6 finds similar results, saying that when activity in the oil industry is high the supplier is busy selling their services and undertaking only incremental changes to their products. However, some companies in our case do have research departments, often focussing on problems which are likely to occur in the future, with the aim of getting one step ahead of competitors.

Successful solutions to customers' problems often provide the reference for future customers. Indeed, several firms said they did not pay for marketing since reputation within the industry is all that matters. However, half of the companies had introduced a new marketing strategy in the period 2012–2014. Several managers stressed the importance of being able to deliver on time and to be flexible should a customer require modifications to the product as the driver of competitiveness. The aim for many firms is to provide a solution which is unique, and where the firm will therefore have a monopoly on future projects.

When the respondents are asked what kind of knowledge is important to stay innovative, most firms regard both engineering and analytical skills as very important (a score of four or five out of five). However, only a few firms value creativity. Engineering skills are considered most important followed by experience and practice and analytical skills while creativity is considered less important. Most managers added that the most important skill their employees can have is curiosity and the ability to work hard to finish a project. Even though they have several collaboration partners, very few explicitly mentioned being able to collaborate and communicate as an important skill. Firm managers tended to emphasise in-house knowledge as most important for being able to come up with new solutions. Approximately 95 per cent of the firms say that the mix is 65 per cent or more in-house knowledge and 35 per cent or less from outside. We also interviewed some oil operator companies, most of which answered the exact opposite: between 15 and 35 per cent internal knowledge and the rest as external knowledge. Several of the subsea firms are part of multinational organisations which means they might also have a lot of resources and in-house competence. The firms depend on collaboration, but they also want to protect their knowledge and technology. Several firms said that the most important way to protect their intellectual property was through contracts with their customers (and in some cases suppliers). Several firms also used other ways of protecting their intellectual property such as patents (11 firms), industrial design (five firms), trademark (eight firms) and claimed copyright (six firms).

Standardisation and the availability of technology are becoming increasingly important, and there has recently been a change in competition within the industry from performance towards cost. Following the fall in oil prices commencing in 2014, cost and cheaper solutions have received much more attention. Several firms mention the positive role of specifications. In particular, governmental safety and environmental regulations have been highlighted as important for innovation and development throughout the whole oil industry in Norway. Some firms say that safety and environmental regulations are the reasons why the Norwegian subsea sector is so competitive on the international market.

The role of collaboration in product development

As in other parts of the petroleum industry, the development of new products in the subsea industry is often undertaken in collaboration with other actors (see also Chapters 2 and 3). Fløysand *et al.* (2012) have previously found widespread collaboration in their survey of subsea firms in Hordaland county. The subsea sector in Rogaland is also highly collaborative. This is reflected in how contracts are set up. Licences for oilfields are typically awarded to groups of multiple oil operator companies. Operators needing subsea services tend to contract with a single subsea firm, but this firm in turn collaborates with other subsea firms during product (and service) development. Quite often, according to the subsea firms, operators pressure for this. Oil operator companies may contract with one firm for equipment on the condition that it uses another firm's component as part of the final product. One informant explained: 'This is how this industry works; you make a product and the customer tells other suppliers they have to use it.' A key to long-term success for a firm is to achieve this position of having a product or component which operators require other (subsea) firms to use. Others say that this drives up costs. Before a product ends up with an operator, three or more other subsea companies have often been involved. Most firms are occasionally direct suppliers to the oil companies and at other times sub-suppliers through other firms' contracts. The firms report a high number of collaborations both vertically along the internal supply-chain of the subsea industry, and also horizontally between competitors.

Figure 4.1 presents the collaboration network between firms in the subsea industry in Rogaland. The density of the network is 0.246, i.e. 24.6 per cent of all possible links in this network are present. The average firm collaborates with seven other firms. The five most central firms in terms of degree centrality are all large firms with more than 200 employees. All have had innovations, and four of five have radical innovations. Firms without innovations are clearly on the periphery of the network. They are all multinationals with headquarters abroad. However, if we also consider linkages outside the subsea sector in Rogaland, two of the top five firms in terms of degree centrality are replaced by other firms, both headquartered in the region.

Figure 4.2 presents the collaboration which subsea firms have with oil operators (top right) and with other oil suppliers outside subsea (top left), to universities and research institutes (bottom right), and to other firms in the subsea collaboration network (bottom left). Most collaboration linkages are with other oil firms while there are fewer linkages to organisations outside the oil and gas industry, indicating that knowledge is mostly sourced from within the oil industry. The firms that most subsea firms collaborate with outside the Rogaland

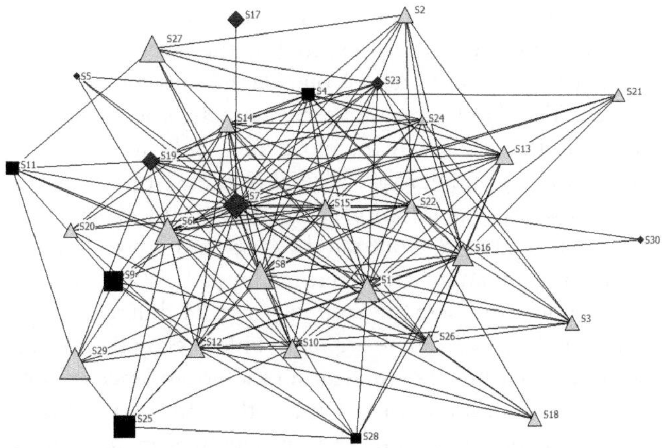

Figure 4.1 Collaboration within the subsea sector in Rogaland.

Note

Size of node refers to the size of the firm. Box nodes are firms without innovation. Diamonds nodes are firms with innovation, and the triangle nodes are firms with new-to-market innovation. (Graph made in Ucinet (Borgatti *et al.*, 2002)).

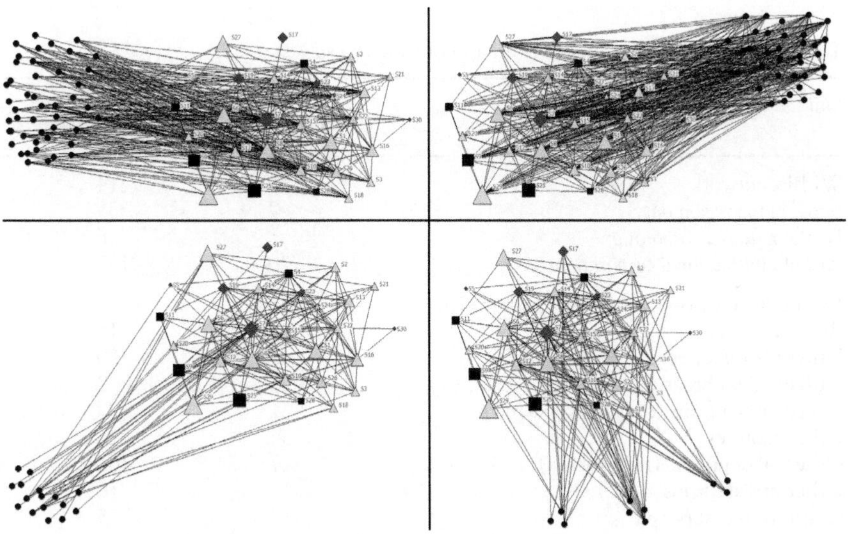

Figure 4.2 Linkages between subsea firms and collaborators.

Note

Linkages from subsea firms to oil supplier (top left). Linkages from subsea firms to other oil operator (top right). Linkages from subsea firms to other firms (bottom left). Linkages from subsea firms to universities, research institutes and other organisations (bottom right).

subsea industry itself are local operator firms (Statoil, Norske Shell, BP) or subsea firms located elsewhere in Norway (Aker Solutions and FMC Kongsberg).

Table 4.3 shows all collaboration partners reported by the interviewed firms in the subsea industry, including links to partners outside the subsea industry. In total, 20 of the firms in the Rogaland subsea industry are part of multinational corporations, and it is likely that they will therefore have links to mother or sister plants elsewhere in Norway and abroad. Several of these firms are international and most state that within-organisation collaboration and resources are very important. Collaboration with actors outside the Rogaland subsea industry is mostly with multinational companies. However, many of these linkages are also to firms in the subsea industry. There are 76 linkages to 21 different subsea firms elsewhere in Norway and abroad. There are more linkages to oil operators than the total number of internal linkages, indicating the dependence on supplier–customer relationships. A substantial portion of other linkages is also to other oil suppliers. Most of the operators and other oil suppliers are also multinational enterprises with offices in Rogaland which are particularly closely involved in the collaboration. Even though subsea firms in Rogaland have several linkages outside the cluster itself, these are mostly to similar industries and their customers, mainly within the region, suggesting that there is a risk of lock-in.

Table 4.3 Collaboration links within and outside the subsea sector

Links in the subsea network	*Number of actors*	*Total number of links*
Within network	30	214
Based in the region only	9	55
Part of national corporation	*1*	6
Part of multinational corporation	20	*153*
From subsea network to external collaboration partners		
R&D institutes	3	15
Universities/colleges in Norway	6	29
Universities abroad	2	2
Other subsea firms (outside Rogaland)	21	76
Oil companies	34	225
Other oil-suppliers (non-subsea)	17	100
Other private firms	15	16
Consulting and people	12	35
Associations (e.g cluster organisations)	7	25
Government	3	14
Total actors based in the region only	9	29
Total national actors	*24*	*112*
(With office in Rogaland)	*(5)*	*(≈20)*
Total multinational actors	87	396
(With office in Rogaland)	*(67)*	*(≈320)*

Oil operator companies are the most common collaboration partners. In total, 28 of the firms report collaboration with Statoil. Several of the subsea firms said that Statoil is the driver of innovation in this industry because it provides funding for several development projects. Concerning other large oil companies such as Exxon, Norske Shell and BP, more than half the subsea firms report collaboration with these. In addition, all the subsea firms said that they monitor the oil companies, and that this gives them inspiration and new ideas.

Concerning collaboration with universities, most firms did not value such collaboration very highly for their innovation ability. They collaborate with universities because they see it as part of their responsibility, also to recruit graduates. Some firms also said that collaborating with universities makes the universities more aware of the particular skills required by the industry. The firms that had formally collaborated with universities in development projects were often frustrated with the time spent on projects, and that they were not allowed to sell or use the product until it was ready and patented. About half of the firms had collaboration with the University of Stavanger and 40 per cent with the Norwegian University of Science and Technology (NTNU).

Challenges for the subsea industry

The period prior to the interviews was characterised by high oil prices and an unprecedented period of growth in the industry. Respondents report that competitiveness is mainly a function of performance and the ability to solve problems rather than of cost. However, after the fall in oil prices, customers and the industry have started paying more attention to costs. This has required subsea firms to cut costs. From 2014 to 2016, most firms have cut wages and laid off employees. It is hard to find exact numbers for how many employees subsea firms in Rogaland have laid off because accounts data are typically reported at the main office. One firm has closed down its plant in Rogaland, although it remains active in Norway, and one firm went bankrupt in 2015. Based on official firm accounts, three-quarters of the firms have lost revenue and reduced wage expenses between 2014 and 2016, almost two-thirds of the firms had a decrease in revenue of more than 20 per cent, indicating that the subsea industry went through a challenging period due to the fall in the oil prices.

The downturn has also resulted in a consolidation of firms within the industry. Many firms have entered into more formal collaboration with other oil firms, either in subsea or other parts of the oil service industry. There have always been mergers, acquisitions and joint ventures in the subsea industry. However, this intensified in 2015–2016 when a quarter of the firms merged, allied or launched a joint venture. This reflects that many firms struggle to remain competitive on their own. For some of these alliances the aim is to reach new markets either by entering the global market, or by entering a new industry.

Discussion and conclusion

This chapter has discussed innovation in the Rogaland subsea industry, focussing in particular on the role of collaboration. Joint projects and other types of collaboration between firms are important for innovation in this industry, as it is in the oil industry overall as described in Chapters 2 and 3. Innovation processes are characterised by close collaboration between subsea firms, and between subsea firms and oil operators and other oil suppliers. The average subsea firm has an extensive network, being linked to seven other subsea firms and to several other oil companies. Large networks are generally assumed to be beneficial in the innovation literature. Even though transaction and communication costs increase, the benefits from interactive learning may more than compensate for these costs (Lundvall, 2013). However, if the networks are too close and rigid, these learning effects may disappear since the network contacts provide too few new insights and ideas.

A potential concern for the subsea sector in Rogaland is that the region has become too specialised and therefore has lost some of its dynamism and flexibility (Martin and Sunley, 2006). There is a risk that 'the local connectedness may become so excessive that fundamental renewal is not on the mind-set and is even heavily contested by local network players' (Boschma, 2015). Furthermore, the industry is heavily dependent on a few central customers, i.e. the oil operator companies. Large firms, like Statoil, have a lot of power. Statoil has been mentioned by several of the firms as important for innovation in the industry. At the same time, the industry is risk averse with strict regulations. With lower oil prices, the oil companies are cutting costs and some of the subsea firms will need to make changes in order to survive. One of the advantages of the subsea sector is their fairly high proportion of engineers whose skills might be relevant in other sectors/industries. This is further discussed in Chapter 11 in this book.

Within a region, the internal network structure is important for knowledge diffusion (Giuliani and Bell, 2005). One strength of the Rogaland subsea network is that the central actors are also innovative firms. This can potentially enable them to share information with actors who are less connected. The potential for information and knowledge flow is high within the network because most actors are connected to more than one other actor. Overall, this is also a highly innovative industry with most firms reporting innovation during the last three years. However, linkages outside the internal network are mostly with firms in the oil and gas industry. The knowledge that comes from outside, both from outside the subsea sector and outside the region, is mostly from within the oil industry. This carries a risk of bringing little new knowledge into the network. Consequently, few firms in the Rogaland subsea industry have successfully managed to move into new markets, and many were heavily affected by the fall in oil prices after 2014. As a result, firms lost revenue and had to downsize; many entered into mergers, alliances or joint ventures with other firms. The future will reveal whether this is an industry that can manage the transition towards new markets in the context of a permanent reduction of activities on the Norwegian Continental Shelf, or whether its high innovation output is mainly geared towards maintaining competitiveness within current markets.

References

Bathelt, H., Malmberg, A., and Maskell, P. (2004). Clusters and knowledge: local buzz, global pipelines and the process of knowledge creation. *Progress in Human Geography*, 28(1), 31–56.

Bathelt, H. and Turi, P. (2011). Local, global and virtual buzz: The importance of face-to-face contact in economic interaction and possibilities to go beyond. *Geoforum*, 42(5), 520–529.

Blomgren, A., Quale, C., Austnes-Underhaug, R., Harstad, A.M., Fjose, S., Wifstad, K., Melbye, C., Amble, A.B., Nyvold, C.E., Steffensen, T., Viggen, J.R., Iglebæk, F., Arnesen, T. and Hagen, S.E. (2015). *Industribyggerne 2015: En kartlegging av sysselsetting i norske petroleumsrelaterte virksomheter, med et særskilt fokus på leverandørbedriftenes eksportsysselsetning. [The builders of Industry 2015: A survey of employees in Norwegian petroleum related businesses, with a special emphasis on supplier industry employees related to exports]* Report IRIS 2015/031. Stavanger: IRIS.

Borgatti, S.P., Everett, M.G. and Freeman, L.C. (2002). *Ucinet for Windows: Software for Social Network Analysis*. Harvard, MA: Analytic Technologies.

Boschma, R. (2015). Towards an evolutionary perspective on regional resilience. *Regional Studies*, 49(5), 733–751.

Fløysand, A., Jakobsen, S.-E., and Bjarnar, O. (2012). The dynamism of clustering: Interweaving material and discursive processes. *Geoforum*, 43(5), 948–958.

Giuliani, E., and Bell, M. (2005). The micro-determinants of meso-level learning and innovation: evidence from a Chilean wine cluster. *Research Policy*, 34(1), 47–68.

Lundvall, B-Å. (2013). Innovation studies: A personal interpretation of the state of the art. Fagerberg, J., Martin. B.R. and Andersen, E.S. (eds.), *Innovation studies: Evolution and future challenges*, 21–70. Oxford: Oxford University Press.

Martin, R., and Sunley, P. (2006). Path Dependence and Regional Economic Evolution. *Journal of Economic Geography*, 6(4), 395–437.

Reve, T., and Sasson, A. (2012). *Et kunnskapsbasert Norge*. [A knowledge-based Norway]. Oslo: Universitetsforlaget.

Ryggvik, H. (2017). Store Norske Leksikon. *Norsk oljehistorie*. [Norwegian petroleum history]. Retrieved from https://snl.no/Norsk_oljehistorie.

Storper, M., and Venables, A.J. (2004). Buzz: Face-to-Face Contact and the Urban Economy. *Journal of Economic Geography*, 4(4), 351–370.

Timmermans, B., and Boschma, R. (201). The effect of intra- and inter-regional labour mobility on plant performance in Denmark: the significance of related labour inflows. *Journal of Economic Geography* 14(2), 289–311.

Additional sources

Aker Solutions (2017). What we do. Retrieved 28.02.2018 http://akersolutions.com/what-we-do/products-and-services.

Petroleumstilsynet (2016). Rapport fra PTIL´s dykkedatabase, DSYS. [Report from PTIL's diving database, DSYS]. Retrieved 28.02.2018 www.ptil.no/om-dykkedatabasen-dsys/category851.html.

5 Cost-cutting as an innovation driver among suppliers during an industry downturn

Jakoba Sraml Gonzalez

Introduction

Between 2009 and 2014, the upstream oil and gas industry on the Norwegian Continental Shelf (NCS) was characterised by unusually high activity and investments. The period ended with a fall in oil prices in mid-2014 and their persistence at a relative low level compared to the previous period for a long time. The fall in prices resulted in an industry downturn and a necessity to lower the costs and ensure profitability of existing and new oil fields. This was a dramatic change from the period 2009–2014. This chapter analyses how suppliers on the NCS reacted to the fall in oil prices since mid-2014 and the impact of cost-cutting initiatives on suppliers' innovation activities and outcomes.

Innovation is often considered as a consequence of 'favourable conditions' – for example, the abundance of resources within a company or in a system (Lundvall, 1992; Nelson and Winter, 1982; Penrose, 1959; Teece *et al.*, 1997). But innovation can also be related to situations of threat in the environment, as Schumpeter (1942) pointed out. Recent studies give examples of how companies can be ingenious at overcoming threats and challenges to come up with new solutions (Gibbert and Scranton, 2009; Keupp and Gassmann, 2013). How exactly and under what conditions companies transform threats stemming from the environment into innovation remains fairly unexplored. Thus, the overall ambition of this chapter is to advance the debate on the dynamics and underlying conditions of innovations in times of resource constraints.

The oil and gas industry as a natural-based resource industry is characterised by volatility due to the link to the commodity market (Williamson, 2012). Periods of activity booms and busts due to fluctuation of commodity prices are common in such industries (Mariscal and Powell, 2014). The actors operating in the oil and gas industry have adapted to the dynamics and learned to take advantage of market fluctuations (Baaij *et al.*, 2011; Dvir and Rogoff, 2009). In fact, cutting costs to increase the efficiency and profitability of offshore operations and development activities has been one of the drivers of technology development in the oil and gas industry (Acha, 2002), particularly in periods of busts with lower oil prices and activity (Gjerde and Ryggvik, 2009). For example, one of the most important innovations in the upstream oil and gas industry – subsea technology – was an industry-wide collaborative initiative to make production systems cheaper, safer and more efficient (Gjerde and Ryggvik, 2009; Ryggvik, 2013, 2015). For this reason, studying how cutting costs drives innovation can also help us to understand the dynamics of innovation in the upstream oil and gas industry.

The upstream oil and gas industry is characterised by two types of activities. The first type is continuous operation for the production of oil and gas where the suppliers provide maintenance and modification of existing infrastructure and processes. The second type are projects for the development of new oil and gas fields, or their expansion where suppliers contribute with their specialised technological capabilities. Such projects, also called Complex Product Systems (CoPS) (Hobday, 2000), are usually one-off customised products for one or a few users produced by a network of different specialised suppliers and other actors (Miller *et al.*, 1995). The characteristic of the CoPS settings importantly influences how the suppliers work with innovation (Davies and Brady, 2000).

The chapter draws on a synthesis across three qualitative studies of suppliers in the upstream oil and gas industry and their innovation activities during the latest downturn. All three are in-depth qualitative studies conducted in different companies in different stages of the industry downturn: initial, severe and consolidation phases. They are based on several data sources: interviews with management and R&D staff, internal documents, multimedia sources and observation. The suppliers in the studies belong to different segments of the upstream supply chain on the NCS and supply products and services to an array of different buyers. The majority of their revenues came from the oil and gas industry at the time of the study. In addition to the results of the synthesis, I present an example (from study 3) of designing a CoPS. The example serves as a showcase of, first, how a supplier works with innovation in a CoPS settings, and second, how the necessity to cut the costs can lead to an innovative outcome.

The structure of the chapter is as follows. I present the existing research on innovation in situations of crises and the empirical context of the study in the following two sections. Then I present the empirical findings and conclude by discussing how threats can lead to innovation.

Innovation and crises

While studies of innovation often argue that companies need supporting and favourable conditions and capabilities to be able to innovate, the literature also emphases the other dimension – that environmental threats, challenging situations or a constraining situation can be conducive to generating novel solutions. Most prominently, Schumpeter (1942) argued that the appearance of an innovation leads to a crisis within the established system which, again, makes room for novelty. Research on innovation in times of financial and economic crisis shows that certain firms continue innovating or even introduce innovations when they experience challenging situations in their environments (Archibugi *et al.*, 2013). For example, Mensch (1975) gave examples of how companies become more risk-prone during recessions since they have no alternative and are forced to think of new solutions when existing markets stagnate.

Organisational studies explain such examples by arguing that when companies face threats they are forced to deviate from routines and come up with new ways of doing things (Williams *et al.*, 2017). They bend existing practices or circumvent the constraining conditions to be able to innovate *despite*, for example, constrained financial resources. Alternatively, companies innovate *because of* the constraining conditions in an attempt to solve a problem associated with them (Gibbert and Scranton, 2009; Keupp and Gassmann, 2013). This does not mean that companies stop progressing when they find themselves in crisis, rather that they can find other creative ways of dealing with the challenges to maintain their position in the industry (Wicken, 1982).

Thus in theory, crises are a good opportunity for innovation (Archibugi and Filippetti, 2011), but in practice there are several limitations, both at systemic and firm levels. From a systemic point of view, Freeman *et al.* (1982) argue that companies are more risk-averse in challenging situations and need a whole system around them to be able to innovate in crises and will thus prioritise process rather than product innovation. At the company level, decisions to sustain R&D in challenging times are often also idiosyncratic and depend on companies' or stakeholders' decisions and previous experiences (Amore, 2015; Kitching *et al.*, 2009; Skålholt and Thune, 2014). Companies tend to satisfice and stick to learned patterns of work when experiencing uncertainty (Daft and Weick, 1984; Nelson and Winter, 1982).

Background: the industry downturn after 2014

The oil and gas industry globally and on the NCS experienced a boom with historically high oil prices, also high costs and investments in the period 2009–2014 (Statistics Norway, 2018). The oil prices suddenly dropped in mid-2014 (as described in Chapter 1). The prices have persisted at a comparatively low level (around 50 to 60 US$ per barrel) since and at the time of writing. The geo-political situation, the behaviour of the actors in the market and the appearance of unconventional hydrocarbon resources set a scenario of persistent low oil prices (U.S. Energy Information Administration, 2018). What seemed like a potentially similar scenario of short-term volatility during the financial crisis of 2007–2008 – sudden collapse and rapid recovery – became a longer downturn in the oil and gas industry.

The new situation in the industry meant that the activities of the oil field operators became less profitable and the overall modus operandi of the oil companies challenged. Consequently, many projects and contracts in the industry have been postponed or cancelled and investments negatively affected (Statistics Norway, 2018). In order for the projects to be realised and profitable, the industry had to lower operation and development costs.

The pre-downturn supplier innovation model

The suppliers in the upstream oil and gas industry serve the operation and development activities. The innovation processes in both types of activities are distributed across different actors (Acha and Cusmano, 2005; Maleki *et al.*, 2016; Perrons, 2014; Shuen *et al.*, 2014). The suppliers contribute to these distributed processes with their technological capabilities. To be able to do so, the suppliers have institutionalised different routines, practices and assumptions, i.e. the supplier 'innovation model'. I discuss the characteristic of the model in the following paragraphs (also see Table 5.1 for the summary of the dimensions and elements of the model).

The suppliers conducted a variety of different types of innovation-related projects. For example, they engaged in projects based on contracts or potential business cases with buyers. They also conducted self-initiated in-house innovation projects or collaborative projects with other industry actors to share the costs or gain complementary knowledge. While the first type of projects were based on contracts and had a secured user and return, the second type were riskier. The projects were funded at least partly internally since they were often conducted upfront in expectation of being taken up by potential buyers.

The nature of selling solutions in the industry contributed to high costs in the decade before the collapse of oil prices in mid-2014. The norm to sell customised or one-off types of project together with a lack of standardisation,

Table 5.1 Dimensions and elements of the established supplier innovation model before the 2014 downturn

Dimension of innovation model	*Elements*
Innovation projects	Internal and in collaboration with buyers or other external actors User involvement – 'Innovation on demand'
Outcomes	Services, products, systems
Purpose of innovation projects	Value for the buyer Increase of internal efficiency
Funding	Own or co-funded by buyers Joint industry projects, alliances
Relation to business logic	Customisation, one-off, lack of standardisation Upfront innovation projects
Institutional aspects	Thorough technology qualification processes Strict focus on Health, Safety, Security and Environment (HSSE) Industry technology standards
Underlying assumptions	Cyclicality (upturn follows downturn) New ideas appear in downturns Necessity for technology Linkage to customer is most important

strict HSSE and technology qualification requirements meant that the suppliers could afford to use resources for adapting every project or solution. Moreover, resource slack in the industry also allowed the suppliers to offer gold-plated and expensive solutions.

The underlying logic of the innovation model among the suppliers was the cyclical nature of the industry. The assumption was that a boom will follow a bust and that the future is known. Such understanding of industry evolution drove risk-prone behaviour and motivated the suppliers to engage in innovation activities also in times of lower activity. They learned that during slow times innovativeness provided a basis for a competitive advantage. Once oil prices had recovered, the industry started investing and buying their services again. When activity in the industry suddenly increased, it was difficult to sustain the focus on innovation because the suppliers were busy selling services and making incremental changes to existing solutions.

The evolving relationship between downturns and innovation activities

The fall in prices and less activity in the oil and gas industry due to a halt in investments and postponement or cancellation of projects meant less demand, reduced revenue and a threat to the suppliers' survival. In addition, buyers started to look for less-costly solutions and often also to replace the suppliers with those who had cheaper alternatives to components or services they needed. The suppliers experienced a double challenge: to cut internal operational costs, and to adapt their solutions to the industry-wide demand to reduce offshore operating and development costs. This in turn put pressure on the established innovation model presented above.

I present three studies of suppliers in three different phases of the downturn in the oil and gas industry, and how they engaged in cost-cutting efforts. I particularly discuss how these efforts had an impact on their innovation activities and how the suppliers adapted them. Table 5.2, at the end of the three presentations, summarises the three studies, the cost-cutting activities of the suppliers and their outcomes.

Study 1: cutting costs to sustain the established innovation model

This study looked at the adaptation of suppliers to maintain disrupted collaborative innovation processes after the first shock of downturn. The suppliers cut costs in internal organisation to deal with the situation; they streamlined their existing offerings in order to eliminate gold-plated or unnecessary 'nice to have' features, lowered their prices as well as adapting their internal innovation-related activities to a situation with fewer resources available. They had to learn how to undertake in-house innovation more efficiently, for example by shortening the engineering hours, reducing the number of engineers, projects and prioritising projects that had a stronger business case. In addition, they adapted procurement activities and reshaped their own supply chains by renegotiating the prices and finding alternatives to costly parts they had previously purchased.

The suppliers wanted to retain the innovation-related activities despite engaging in retrenchment. Moreover, some – particularly the EPC (Engineering, procurement, and construction) companies in the study – proposed new solutions to deal with the low prices that would not target the short-term need for cheaper, but that would be cost-effective also in the long run and thus target the profitability and not only the necessity to spend less. According to the suppliers, this was the most optimal way of dealing with the necessity to bring down the operating costs – targeting the design. Nevertheless, they experienced that while the buyers were interested in lowering the costs, they prioritised cheaper and less-costly solutions.

Study 2: considering changes to the innovation model

The second study analysed how a specialised hi-tech supplier considered technology choices when the survival of the company was threatened, and the existing technology strategy became unsustainable during the downturn. The supplier downsized and reorganised R&D besides cutting costs in the whole organisation. It also adjusted the pace of the R&D projects by slowing down the non-core projects and consolidating similar ones. Nevertheless, the responses were not enough in what became a prolonged period with lower demand, lower revenues and the consequent lower R&D funding. The changes made in cutting costs did not enable them to sustain R&D activities in the long run in an environment with severely constrained financial resources. The buyers prioritised competitors with less-costly technological solutions and the company in this case had to lower drastically the prices of their high-quality services. As a response, the hi-tech supplier engaged in discussions on how to adapt more radically and what specific changes were required to the innovation model. Nonetheless, the company was reluctant to make more radical changes to their model apart from incremental modifications due to the uncertainty about the persistence of the market conditions that triggered cutting costs in the first place.

Study 3: adapting and expanding the innovation model

The third study looked into how a supplier adapted its position in relationships with buyers in a consolidation phase of the downturn, and how this affected its innovation activities. Similarly to the first two studies, the supplier engaged in streamlining and reorganisation of the R&D organisation. It also engaged in more radical changes to its innovation activities since the situation in the industry had deteriorated over time. This meant not only cutting the costs, but also reorganising its structure in order to fit it with the new demand. The company introduced changes to the innovation management, particularly to its value creation aspect with more profound effects on innovation activities. The

Table 5.2 Summary of the three studies, cost-cutting activities and outcomes

Study	*Phase in downturn*	*Cost-cutting activities*	*Outcomes*
Study 1: 10 suppliers in different segments	*Adapting to the first shock of downturn* (November 2015–August 2016, with one additional interview in December 2016)	• retrenchment (costs, projects, downsizing, streamlining solutions) • proposing concepts for less costly solutions	• increased internal collaboration • work process changes • some: external alliances • some: diversifying • some: internal R&D initiatives for new less-costly concepts
Study 2: 1 supplier	*Assessing more radical response to downturn* (March 2016–February 2017)	• retrenchment (costs, projects, downsizing) • revising of the innovation strategy	• work process changes • digitalisation • new R&D organisation • optimisation of technology use • stronger focus on risk calculation
Study 3: 1 supplier	*Shaping new innovation processes and activities* (April 2017–October 2017)	• retrenchment (costs, projects, downsizing) • marketing less costly solutions	• work process innovation • digitalisation • change of innovation strategy • new R&D organisation • new transformative innovation projects • diversification • stronger focus on risk calculation

institutionalisation of changes occurred because the company saw the necessity to be more proactive in dealing with the crisis. Furthermore, the company established a more ambitious innovation strategy which targeted not only coming up with solutions for buyers in established markets, but also for new markets.

Cross-case comparison

The three studies show how cost-cutting was necessary for the suppliers to adapt and to sustain their innovation activities during the downturn. All companies changed their work processes to increase efficiency. This covered operation processes and processes related to innovation activities, since the two aspects are often intertwined in a CoPS setting. One salient issue was digitalisation.

Looking across the studies, cost-cutting initiatives and their impact on innovation activities became more radical over time. The change was mainly in the scope of the innovation activities. The companies in the first study adapted the focus of their innovation activities, but sustained the established model of offering conceptual or ready-made solutions to their buyers. The company in the second case was forced to rethink the innovation model due to its unsustainability under cost-cutting and demand pressures. Finally, the company in the third case adapted its own innovation activities and expanded its scope beyond the pre-downturn model.

While the studies are not comparable, the findings suggest that the response to the crisis went from mere reorganising and lowering the costs of existing ways of conducting engineering and innovation-related activities to considering and introducing new types of innovation efforts and adapting the internal model. Only cutting costs by stopping activities or removing expensive components was not enough; they also had to consider more substantial modification and reorientation of their solutions and innovation activities.

On the one hand, the change was due to the deterioration of the economic conditions in the environment, and the consequent inability of the companies to sustain the existing model over time. On the other hand, the change was supported by a changing perception of the downturn over time. What was supposed to be a well-known scenario of a short-term crisis according to many of the informants turned out to be a prolonged period that challenged some of the assumptions about the innovation model. For example, the hi-tech supplier in the second study experienced three similar periods in the industry before and accumulated experience on how to deal with them. Despite having adapted the innovation strategy according to the lessons learned from previous downturns, the latest one still challenged it. This time the company reported less certainty about future demand due to mixed signals from their buyers. The expectation of cyclicality gave certainty about activity in the industry, but the peculiarity of the latest downturn created uncertainty about the pace and direction of demand for their services, and finally also about their innovation strategy.

The impact of cost-cutting on innovation outcomes: an example of a FEED project

In this section, I present a case of a front-end engineering design phase (FEED) project that a supplier executed together with the buyer. The case exemplifies the interactive process of cutting costs and coming up with novel solutions as the consequence. In particular, it exposes the related challenges of overcoming established routines of working with buyers, and the supplier's internal challenges of actually coming up with creative ideas and innovation solutions.

One particular aspect of work that became even more crucial during the crisis and enabled lowering the costs in the specific case of development of new fields was the concept work in the FEED phase. The development of fields is a step in a field lifecycle, which consists of defining the characteristics of the field and its elements. The FEED consists of planning the project before a final decision about the sanctioning of a new field development project is made. The components of a project, its design and the costs involved are specifically defined in this phase. In this sense, the FEED is the primary factor in defining the economics of a project and the profitability of the project outcome. The fall in the prices and their persistence at lower levels puts pressure on the profitability of many of the planned projects. Namely, the break-even price – when a field is profitable – dropped significantly. This in turn meant that the industry had to evaluate more thoroughly how its capital goods are conceptualised and designed to enable their profitability in a low oil price environment.

The planned project for the CoPS on the NCS in this example was in the planning phase with a concept being defined when the oil prices started to fall. The dramatic change in the absolute levels meant that the profitability of the project was no longer guaranteed. Nevertheless, the project had a high strategic attention from the buyer which meant that they opted for revising the concept instead of cancelling or postponing the project. For the supplier involved in the concept design work, the case also had importance since, first, it would showcase their capabilities, Second, the supplier required sufficient revenue to secure its operation when many other projects were cancelled and there was less demand for its products and services. Therefore, the project had a high priority from the buyer's and supplier's side to be deemed profitable in a low oil price environment, and eventually sanctioned.

The cost-cutting activities started with the reduction of the prices from the suppliers' side. For example, they cut the margins, replaced the expensive components where possible and streamlined their own solutions. Yet, the cost of the solution was still too high for it to be profitable and acceptable for the buyer. Consequently, in an attempt to really lower the costs, the supplier challenged the buyer to contribute to the cost-cutting process by adapting their technical requirements, to which the buyer eventually agreed. The supplier and buyer ended up in a collaborative process of exchanging evaluations of costs where consensus about what to cut was achieved. In this process the buyer adapted its technical requirements and the supplier contributed by internally designing different solutions. For the supplier, this meant that its engineers had to come up with new solutions that would satisfy the new requirements in a cheaper way. For example, when the buyer allowed the supplier to reduce the number of a specific component from two to one, the supplier consequently was not only able to reduce the cost by using less equipment, but also to either reduce it or use it for another purpose.

The design process was complex for the supplier because, first, the engineers involved were challenged to be creative and think differently about how to design solutions that retained the quality at a lower cost. The supplier introduced a work-process innovation to support them. This was a new approach to engineering that consisted of cross-boundary work of identifying costs and designing solutions that targeted their lowering. FEED work is by definition done by engineers with a holistic overview of the CoPS. The new engineering approach institutionalised the collaboration between engineers in different disciplines and components specialists even more. The outcome was an increased sharing of competences across disciplines and departments.

Second, there was uncertainty about what reducing costs meant in practice. The supplier had to challenge the buyer about their requirements and come up with new ones. This was not the practice in an industry where the buyer has more bargaining power relative to the supplier since they have the final decision about the contracts and design. In this process, the buyer's new requirements served as guidance for the engineers in actually deciding on a new design. Without a consensus the suppliers had a much larger set of possible options on how to come up with a design with lower costs. To reach a consensus in practice, the buyer and supplier co-funded a team of process facilitators that eased the interaction and enabled the new requirements and the necessities of both to be expressed in a tangible way. The final outcome of the interactive process was an innovative CoPS design with significantly reduced costs for the buyers and a much lower break-even price than at the start, but still with all the necessary functionality as with the original, more expensive, design.

Innovative outcomes in downturns

The above example shows that cost-cutting processes can lead to an innovative outcome, are triggered by the necessity to adapt to a threat, and are enabled by the collaboration between the buyer and the supplier. According to the supplier, the cost-cutting process led to an innovative outcome because both actors wanted to arrive at a consensus and move on with the project. This was not possible before the downturn or in the early phases because there was neither necessity nor urgency to make an effort to reduce the costs of the design. The established patterns of collaborative work were repeated to avoid engaging in time-consuming and complex changes, and also because they expected the situation to return so that there would be no need to do things differently. The first study exposes a similar condition but a different outcome. Many of the suppliers proposed working on new concepts and solutions, but did not have the commitment from the buyers to adopt them which in turn limited the deployment of the solutions albeit that these were cost-effective.

The case studies also show that the collaborative process to come up with an innovative outcome had an unintended consequence for the supplier in the form of work-process innovation. The new process where internal engineers from different disciplines collaborated with the buyer's engineers in a creative action was introduced to help the engineers in dealing with the complex design task. The new work approach allowed them to make the step from just reducing the price of the solution to actually designing a solution that is inherently less costly but with the same functionality. Similarly, the suppliers in other cases also engaged in changes to their work processes to increase efficiency, for example, by joining alliances (study 1), internal optimisation of the use of technology assets (study 2) and digitalising some tasks (studies 2 and 3). All these were unexpected outcomes of trying to deal with the threats to their innovation activities.

Finally, this shows that internal retrenching alone was not enough for coming up with an innovative outcome nor, for the suppliers, was working alone on their own concept sufficient. Rather, the innovative outcome came from a holistic approach where several actors within the supplier and buyer organisations collaborated, also across boundaries.

Discussion and conclusion: how threats lead to innovativeness

This chapter has discussed how threats lead to innovativeness drawing on three studies of suppliers in the upstream oil and gas industry during the downturn following the fall of prices in mid-2014. The chapter has presented a synthesis of three empirical studies of suppliers in different segments of the upstream petroleum value chain and focused especially on how cost-cutting activities, as a response to the downturn, are linked to innovation activities.

The results indicate that the companies engaged in individual actions to cut the costs, but also took part in collaborative cost-cutting initiatives aimed at changes of systemic nature. The urgency to adapt during the downturn was directed towards changing broader units than just individual components or technologies. Changes solely involving individual solutions or companies did not allow the necessary lowering of the costs, as shown in the example of FEED.

Furthermore, I found that the cost-cutting processes were collaborative also in another way; the actors (buyers and suppliers) were interdependent also in individual initiatives. The suppliers were willing to sustain the cost-cutting initiatives with more radical individual solutions, but they depended on the buyers' requirements in shaping a consensus about what a more radical solution should even be. The findings are in line with previous studies of technical change in times of downturn in the oil and gas industry which show that while the necessity to increase efficiency due to cost pressures started as several individual actions, the technology change occurred at a systemic level as collaboration among several actors (Gjerde and Ryggvik, 2009; Ryggvik, 2013).

The two findings expose the distributed and interactive nature of innovation processes in the upstream oil and gas industry (Acha and Cusmano, 2005; Perrons, 2014) where actors depend on each other in shaping innovative outcomes. The suppliers depend on buyer's decisions and buyers depend on the suppliers' technological capabilities. Under threat, these processes showed characteristics of a holistic approach where actors crossed boundaries and depended on each other to come up with significant innovative outcomes and not merely pure reduction of costs of existing solutions. What they also showed was the challenging aspect of these processes. Tensions arose together with the necessity to change due to organisational issues like coordination, but also deeper structural issues like unequal power positions in the collaborative relationships.

The findings about the adaptation of the suppliers to the downturn show how the suppliers initially introduced process changes to sustain the established innovation model *despite* the challenges in the environment. The process changes eventually played an important role in achieving the outcome. This was a necessary step in a situation of economic constraints. At the same time, the process changes also enabled innovativeness and the design of new solutions to address cost-cutting pressures. The first study – as a representation of the first part of the crisis when there was uncertainty about how to sustain the old innovation model – shows how they still engaged in innovation activities *despite* the crisis. But the second and third studies in the later stage of the crisis – when a more radical need to cut costs appeared and the old situation could not be sustained – show how the companies still had to be innovative *despite* the constraints internally and *because of* the pressure to further cut costs in the industry. While innovative products or services comprised the most important outcome for the suppliers, work-process innovation was the unexpected outcome which eventually enabled them to engage in designing new cost-effective solutions. The literature on innovation under threat argues that innovation *despite constraints* and *because of constraints* characterise different innovation processes (Gibbert and Scranton, 2009; Keupp and Gassmann, 2013), but the findings in this chapter show that the two modes can co-exist. The reason for the co-existence is that the companies experience different sources of threats and different degrees of necessity to address them during an industry downturn.

Overall, this study indicates how a threat can lead to innovativeness by creating the necessity for companies to change in order to survive. The magnitude of the threat is of significance since the appearance of the 'sense of urgency' functions as a driver to engage in not only adapting existing innovation activities, but also to come up with significant changes – either in process or product. These changes enable a continued sustainability of the operation and the position of the company in the market.

References

Acha, V. (2002). *Framing the Past and Future: The Development and Deployment of Technological Capabilities in the Upstream Petroleum Industry*. PhD Thesis, University of Sussex.

Acha, V. and Cusmano, L. (2005). Governance and co-ordination of distributed innovation processes: patterns of R&D co-operation in the upstream petroleum industry. *Economics of Innovation and New Technology*, 14(1–2), 1–21.

Amore, M. D. (2015). Companies learning to innovate in recessions. *Research Policy*, 44(8), 1574–1583.

Archibugi, D. and Filippetti, A. (2011). *Innovation and Economic Crisis: Lessons and prospects from the economic downturn*. Abingdon: Routledge.

Archibugi, D., Filippetti, A. and Frenz, M. (2013). The impact of the economic crisis on innovation: Evidence from Europe. *Technological Forecasting and Social Change*, 80(7), 1247–1260.

Baaij, M., de Jong, A. and van Dalen, J. (2011). The dynamics of superior performance among the largest firms in the global oil industry, 1954–2008. *Industrial and Corporate Change*, 25(3), 1–36.

Daft, R. L. and Weick, K. E. (1984). Toward a Model of Organizations as Interpretation Systems. *The Academy of Management Review*, 9(2), 284–295.

Davies, A. and Brady, T. (2000). Organisational capabilities and learning in complex product systems: towards repeatable solutions. *Research Policy*, 29(7–8), 931–953.

Dvir, E. and Rogoff, K. S. (2009). Three Epochs of Oil. NBER Working Paper No. 14927. National Bureau of Economic Research.

Freeman, C., Clark, J. A. and Soete, L. (1982). *Unemployment and technical innovation: a study of long waves and economic development*. London: Pinter.

Gibbert, M. and Scranton, P. (2009). Constraints as sources of radical innovation? Insights from jet propulsion development. *Management & Organizational History*, 4(4), 385–399.

Gjerde, K. Ø. and Ryggvik, H. (2009). *On the edge, under water*. Stavanger: Wigestrand.

Hobday, M. (2000). The project-based organisation: an ideal form for managing complex products and systems? *Research Policy*, 29(7–8), 871–893.

Keupp, M. M. and Gassmann, O. (2013). Resource constraints as triggers of radical innovation: Longitudinal evidence from the manufacturing sector. *Research Policy*, 42(8), 1457–1468.

Kitching, J., Blackburn, R., Smallbone, D. and Dixon, S. (2009). *Business Strategies and Performance During Difficult Economic Conditions*. London: Department for Business, Innovation and Skills.

Lundvall, B.-Å. (ed.). (1992). *National Systems of Innovation: Towards a Theory of Innovation and Interactive Learning*. London: Pinter Publishers.

Maleki, A., Rosiello, A. and Wield, D. (2016). The effect of the dynamics of knowledge base complexity on Schumpeterian patterns of innovation: the upstream petroleum industry. *R&D Management*, in press.

Mariscal, R. and Powell, A. (2014). Commodity Price Booms and Breaks: Detection, Magnitude and Implications for Developing Countries. IDB Working Paper Series No. IDB-WP-444. Department of Research and Chief Economist, Inter-American Development Bank.

Mensch, G. O. (1975). *Stalemate in Technology. Innovations Overcome the Depression*. Cambridge, MA: Ballinger.

Miller, R., Hobday, M., Leroux-Demers, T. and Olleros, X. (1995). Innovation in Complex Systems Industries: the Case of Flight Simulation. *Industrial and Corporate Change*, 4(2), 363–400.

Nelson, R. R. and Winter, S. G. (1982). *An Evolutionary Theory of Economic Change*. Boston: Harvard University Press.

Penrose, E. (1959). *The Theory of Growth of the Firm*. London: Basil Blackwell.

Perrons, R. K. (2014). How innovation and R&D happen in the upstream oil & gas industry: Insights from a global survey. *Journal of Petroleum Science and Engineering*, 124, 301–312.

Ryggvik, H. (2013). *Building a skilled national offshore oil industry: The Norwegian experience*. Oslo: NHO.

Ryggvik, H. (2015). A Short History of the Norwegian Oil Industry: From Protected National Champions to Internationally Competitive Multinationals. *Business History Review*, 89(1), 3–41.

Schumpeter, J. A. (1942). *Capitalism, Socialism and Democracy*. London: Taylor and Francis, 2006.

Shuen, A., Feiler, P. F. and Teece, D. J. (2014). Dynamic capabilities in the upstream oil and gas sector: Managing next generation competition. *Energy Strategy Reviews*, 3, 5–13.

Skålholt, A. and Thune, T. (2014). Coping with Economic Crises – The Role of Clusters. *European Planning Studies*, 22(10), 1993–2010.

Teece, D. J., Pisano, G., and Shuen, A. (1997). Dynamic Capabilities and Strategic Management. *Strategic Management Journal*, 18(7), 509–533.

Wicken, O. (1982). Småbedrifter og primitivisering. F. Sejersted (Ed.), *Vekst gjennom krise. Studier i norsk teknologihistorie*. Oslo, Norway: Universitetsforlaget.

Williams, T., Gruber, D., Sutcliffe, K., Shepherd, D. and Zhao, E. Y. (2017). Organizational Response to Adversity: Fusing Crisis Management and Resilience Research Streams. *Academy of Management Annals*, 11(2), 733–769.

Williamson, J. G. (2012). Commodity Prices over Two Centuries: Trends, Volatility, and Impact. *Annual Review of Resource Economics*, 4, 185–352.

Additional sources

U.S. Energy Information Administration. (2018). Short-term energy outlook. Retrieved 02.2018 www.eia.gov/forecasts/steo/report/prices.cfm.

Statistics Norway (2018). Investments in oil and gas, manufacturing, mining and electricity supply. Retrieved 23.02.2018. www.ssb.no/en/energi-og-industri/statistikker/kis/kvartal.

6 Norwegian rig service industry

Innovations in contractual relations

Petter Osmundsen

Introduction

Oil companies are struggling to develop new projects in Norway. The problem started before the fall in the oil price. Even at a high oil price, project profitability failed after years of rig rate increases that commenced about 2002 and culminated in 2013, with day rates for floaters increasing from around 110,000 dollars in 2002 to around 500,000 dollars in 2013. The crisis in the petroleum industry called for dramatic solutions. While Chapters 2 and 3 analyse innovations on a sectoral level and the role of suppliers supporting innovations, this chapter focuses on innovations in contractual relations implemented by oil companies towards the supply industry to curb drilling costs. In an effort to reduce rig rates, the oil companies investigated whether new ways of organising the relationship between themselves and the rig contractors – including changes to risk-sharing and ownership – could increase the supply of units at an acceptable cost. In the peak years, the scarcity of rigs, combined with a considerable lengthening in contract duration, led to a number of interesting examples of innovation in contractual and organisational patterns for drilling. These include new types of incentives in drilling contracts, small oil companies joining forces to establish a rig consortium, and vertical integration where oil companies own rigs.

Examples and cases presented in the chapter are taken from the Norwegian continental shelf (NCS). Since all petroleum provinces have experienced rising rig rates, since major players among oil companies, rig companies and oil services are present on the NCS, and since the contracts for rig hire mostly follow an international standard, the findings in the chapter are likely to have global relevance, i.e. the chapter portrays international trends in rig procurement. To some extent, the changes can also be seen as the Norwegian oil industry adjusting to particular international trends for mature provinces. One adjustment in that respect are changes in contracts and organisational pattern to better fit the entry of new oil companies on the NCS. Many of these are smaller companies with a lower ability to carry risk and with a limited technical staff. They demand a larger set of services from the supply industry, both in terms of technical services and carrying of risk. This is a change we have seen in other more mature areas, for example in the UK.

A distinction can be made between fixed and mobile drilling units. This chapter addresses mobile units or rigs. The licensees have normally always owned the drilling units permanently installed on platforms. Crews are hired from drilling contractors, and maintenance is usually also contracted out. The mobile rigs operating on the NCS are owned by a rig contractor and chartered in to conduct drilling operations. Such units can be classified according to two main categories – jack-ups standing on the seabed, and semi-submersible floaters. Drill ships also exist, but only one of these is found on the NCS.

Innovation in the oil service industry is in this chapter portrayed by some pivotal cases of industrial restructuring. To analyse the cases a broad scientific approach is applied. The chapter builds on general economic incentive and contract theory, see Hart (1995), Hillier (1997), Milgrom and Roberts (1992); more specific research in the drilling field, such as carried out by Corts (2000), Corts and Singh (2004), Osmundsen *et al.*, (2015, 2010, 2012, 2006), and Osmundsen, Sørenes and Toft (2008, 2010); and literature on petroleum contracts and organisation, e.g. Moomjian (1999), and Osmundsen (2010, 2013), Osmundsen *et al.*, (2008); literature on investment cycles in the petroleum industry, see Osmundsen *et al.*, (2006, 2007); and government appointed studies on rig activity and increased oil recovery, e.g. Rig Commission (2012) and Improved Recovery Commission (2010). In addition, I have had a series of meetings and conversations with specialists closely involved with decisions related to rig procurement, including technical and organisational conditions, legal aspects and taxation. I have also had access to rig contracts used on the NCS.

It should be noted that the subject is complex with no clear and simple answers to the issues involved. A multitude of different company and contractual structures exists for rigs, production ships and so forth, worldwide depending on the tax regime, ownership systems for licences, market conditions and so forth. Theoretically, players often weigh up different considerations when choosing contractual and organisational solutions. Empirically, successful companies applying different solutions can be found side by side in a sector. A corresponding wide array of approaches to rig procurement exists internationally. A number of interesting general insights of a conditional nature can nevertheless be found. On the basis of theory and available empirical considerations, something can be said about the conditions where specific organisational and contractual solutions are best suited.

New type of rig contractor

The NCS has been developed by big oil companies, and contractual templates are tailored accordingly. New companies have other requirements, including risk-sharing. In drilling, the oil companies bear the main oil price, foreign exchange and production risk. The issue of contracts where the contractor bears a larger share of the risk is making itself felt in Norway through the establishment of new and often smaller oil companies. These lack the same ability to carry financial risk as existing companies on the NCS and depend on sharing risk with their contractors. In exchange, the latter get to share in the upside of the projects (see Chapters 2 and 14).

The company Aker Exploration was created to meet the need for greater contractual diversity. It offered rigs for exploration drilling on the NCS. In exchange for bearing the whole drilling cost, Aker Exploration obtained a share of the oil or gas field in the event of a discovery. This represents vertical integration – a form of enterprise which is both an oil company and a provider of rig services – and is an interesting innovation on the NCS. In addition to securing sorely needed rigs tailored for Norwegian requirements, it represents a contractual novelty which reduces cost and risk for the oil company. This should be particularly suited for new players on the NCS.

Compared to traditional rig contractors, however, this model makes bigger demands on the ability to bear risk and breadth of expertise – which must also embrace energy markets and reservoir conditions. The challenge is to avoid the problem of adverse selection, or a position where oil companies exploit their better reservoir knowledge (asymmetric information) to offer co-ownership only for the least attractive acreage and to tackle the other blocks on their own account. However, adverse selection is a general problem for companies seeking to farm into oil fields. Another problem involves overcoming challenges related to a second-hand market for licences on the NCS which at times can be illiquid. The big companies on the NCS have traditionally demonstrated little interest in farming out licences. On the basis of ambitious production and reserve goals, they have often been keener to swap licence holdings than to sell them. Exchanging rig capacity for a licence interest could therefore prove a demanding issue since Aker's business model here depends on convincing all participants concerned in the relevant licences. This normally involves time- and resource-intensive negotiations. With good help from a tight rig market, however, the company succeeded in securing deals. For example, Aker Exploration entered into an agreement on taking over 30 per cent of the oil company ENI's sole holding in production licence (PL) 259 in exchange for providing the Aker Barents rig. Aker Exploration also acquired ENI's 55 per cent interest in PL 256 (DN.no, October 17, 2007). In 2009, Aker Exploration merged with Det Norske to become Det Norske Oljeselskap (e24.no, 7 November, 2009).

Linking licence transactions with rig provision would represent a socio-economic benefit if it increased the liquidity of the licence market. Generally speaking, this liquidity appears to have increased on the NCS, and Aker Exploration's timing was good. One challenge would have been that liquidity varies over time, which was perhaps one reason for the merger. The combination of rig and own licence interests seems more flexible since the need to swap rig provision with licence interests has declined. To some extent it also opens the way for licence swaps.

Swapping licence interests for drilling capacity represents a further development of farm in/farm out arrangements. Such arrangements could be prompted by a lack of rig capacity or budgetary constraints on exploration funding by the oil company undertaking the farm out. This operation is particularly relevant for new entrants on the NCS which have not built up their own production. Farm in/farm out deals have previously been agreed between the oil companies. Aker Exploration and Aker Drilling are examples of a combined solution for the rig contractor and an oil company using shared equity interests to implement such arrangements. Certain other rig contractors internationally also offer solutions of this kind. These require the contractor to expand its expertise base beyond rig operation, and to acquire knowledge of reservoirs in order to deal with the new risks and opportunities it faces.

A relevant question is whether the examples of vertical integration where an oil company secures a rig or a contractor seeks licence interests and is driven primarily by short-term economic conditions represent a sustainable and fundamental business strategy. The shortage of rigs was driving special solutions. In addition, very high rates make rig ownership tempting for oil companies as a supplementary business activity. However, it remains to be seen whether these types of strategy survive an economic downturn with a buyer's market for rig services. Oil companies presumably do not normally own rigs for one main reason. Several hazards exist and the risk is high since it cannot normally be spread to the same extent achieved by big international oil companies through part-ownership of many different licences. Small oil companies lose flexibility by owning rig capacity, and it is uncertain whether they have the financial muscle to retain such ownership over time. However, risk exposure will be affected by the contracts employed, and new solutions are also conceivable here. If rig slots can be sold in many fields with different geological structures, technical portfolio diversification will be achieved. The precondition then is that the activity can achieve a certain scale and spread. Another common objection to oil companies owning rigs is that this does not form part of their strategic core (focusing strategy) – one cannot be successful in everything. Over time, an oil company's own rig fleet would not normally match its licence portfolio, and it could quickly end up in the position of providing a rig as a contractor to customers it competes with as an oil company. An international player will face geographical mismatches, and moving rigs is expensive. An oil company will also normally have varying demand for rigs. The big international players do not normally want to sit with continuous rig exposure through ownership when their drilling requirements vary sharply both over time and in geographical spread. Overall, therefore, many considerations argue for outsourcing here. The exception could be conditions where a long-term drilling requirement with a specific type of rig apply (see Chapter 7).

New contracts

Rig contractors and oil service companies must be urged to formulate contracts suitable for new small players on the NCS. Such contracts would need to involve different risk-sharing arrangements compared to current contracts. The latter are influenced by the fact that players on the NCS have been large international companies with a substantial ability to bear risk and with a high level of expertise in the management of drilling operations. Circumstances are different for a number of new entrants on the NCS. They will want to shift more risk to their contractors, and depend much more on purchasing external expertise. To meet this demand, contractors must extend their expertise base and develop suitable systems for risk management. However, risk exposure must be carefully reconciled at all times with the contractor's ability to bear risk. Research from the Gulf of Mexico shows that turnkey contracts are primarily used for exploration wells drilled from jack-ups in shallow water, and that the companies employing such contracts are small players with limited experience and financial strength (Corts, 2000). Exploration wells in the North Sea for some of the new companies on the NCS should fit that description. Well intervention is another possible example. However, establishing such contracts requires drilling contractors willing to bear the increased risk and to expand their scope of activity and range of services. This seems to be case for only a limited number of existing contractors. A clear distinction exists, for example, between drilling contractors and oil service companies, and few appear particularly willing to bear reservoir and oil price risk.

Intermediate solutions are possible – not a single turnkey contractor for drilling, but at least fewer suppliers because more drilling services are delivered by the same oil service company. This simplifies procurement and management processes for the oil company. It also permits wider use of incentive agreements because a contractor delivering more services gains better control of the drilling process. The collaboration between Pertra and Halliburton indicates that new procurement models of this type could boost value added. A trend towards coordinated deliveries should also be interesting for international oil companies since the benefits – more coordination and reduced transaction costs – seem comparable with the coordination on the contractor side seen in development projects with the introduction of engineering, procurement, construction and installation (EPCI) contracts. However, the advantages of greater coordination by contractors must be weighed against the drawbacks of reduced competition – in practice, few companies can offer such a broad range of services – and less specialisation.

Cost increases in the oil industry are a significant problem. Decentralised contract structures in this area can mean sub-optimisation. While very strong incentives (competitive rates) can be optimum at the project level, they can drive up offshore costs as a whole. Thus, seen from the perspective of government, the welfare effects of new contract forms have to be weighed up with the possible efficiency benefits of stronger incentives balanced against a rise in costs. While large oil companies will internalise much of the cost increase at the overall continental shelf level, and thereby have virtually the same interests as the government, those with a small portfolio on the NCS will, by and large, place emphasis on incentive considerations. The supplementary incentives in oil service contracts are reportedly profitable for the individual licensee. When evaluating profitability at the continental shelf level, however, possible contagious effects in the form of increased rates in competing licences must be taken into account. But the supplementary incentives represent such small amounts that they are not a significant problem. Moreover, innovative thinking should be welcomed in a contract area which has been very conservative.

New oil companies – rig consortium

New companies on the NCS were particularly affected by the rig shortage (see Chapter 2). It could be more difficult for them to gain access to rigs since the rig contractors primarily want to deal with big customers who place large, long-term charters. This could create problems for newcomers in fulfilling government-imposed drilling commitments specified in work programmes. On top of that, of course, are the expectations and demands of impatient shareholders.

Seven small companies joined forces to charter a rig through a consortium led by DPT (now AGR). These players were largely located on the same site, the Veritas building in Stavanger. That has permitted the creation of a joint safety centre for drilling operations which is operated on a 'shift' system by the company using the rig at any given time. In other words, the benefits of scale and shared location accrue here to the operator. The new players have otherwise gone further than the big companies in outsourcing rig management. Another player in the rig consortium field is Rig Management Norway.

Small oil companies want to share risk differently with their contractors and to use other contract formats, for example with the contractor bearing a number of consequential costs in the event of downtime, or sharing the production and oil price risk. An obstacle here is that contractors normally lack the ability or willingness to accept this type of risk. Their argument is that the new companies are not prepared to pay for all the responsibility they want to transfer. Cases exist from the UK where this type of risk transfer to the contractors led to renegotiation of the contract – the contractor was unable to bear the risk. Similar experiences have been encountered in development projects on the NCS. The lesson is that incorporating more risk than the contractor can bear would not be appropriate. With some exceptions, contractors basically do not want contracts linked to production volumes or sales since they are not willing to accept reservoir or price risk. On smaller fields, however, it could be the case that contractors are better suited to bearing this type of risk than newly established oil companies.

One implication of the lack of ability and willingness to bear risk in the contractor chain is that not all oil companies are suited for pursuing major field developments. In a number of cases it would be preferable to sell interests to companies with sufficient ability to execute the project. Developments on the NCS often demand much from the operator's organisation in the form of expertise and capacity. A number of jobs in such areas as project follow-up are not appropriate for outsourcing. They form part of the strategic core in the oil companies. Furthermore, such projects make big demands on the capital adequacy and liquidity of the licensees – cost overruns and delays can be difficult to cope with for new players. Nevertheless, there are indications that the market for bearing risk develops in the contractor sector in order to meet this new demand.

New rig types

Statoil moved from a position where it stopped the chartering process for rigs in 2008 and adopted a more active procurement approach which included the design of new rig categories and assessing rig ownership. An enquiry for chartering a rig had been issued by StatoilHydro in the summer of 2008, covering both semi-submersibles and jack-ups, commencing in 2012. The company requested updated bids from contractors and secured a reduction in the rates offered. It nevertheless decided to terminate the process because the rates were considered too high.

In collaboration with international contractors, Statoil developed completely new rig concepts for use in improved recovery and on new fields. In May 2011, it awarded a contract to Songa Offshore for two new Category D units to be used on the NCS. A total of four such rigs have been ordered by the company from Songa. The goal is that the new type will do the work 20 per cent more efficiently than conventional rigs. Challenges arose in relation to Songa's capital adequacy, and problems with yards and delays were reported (DN.no, 26 August, 2013). Statoil was the biggest contributor to what is intended to be the final rescue of financially distressed Songa Offshore. Statoil, creditors and owners helped to save the company from liquidation (DN.no, 25 November, 2013). Statoil's contribution took the form of raising rates for the four rigs by US$300 million in all. US$100 million will be repaid if it exercises all eight options for lengthening the four rigs.

In April 2012, Statoil entered into a contract with Aker Solutions for a new Category B rig concept. This unit is designed to conduct various types of well interventions with wire-lining and coiled tubing, and for through tubing rotary drilling of side-tracks. Aker Solution's share price reacted negatively to the signing since income was fixed while costs were uncertain. The parties agreed in June 2013 to cancel the charter for the new rig category. As I understand it, disagreement had arisen over sharing the cost of maturing the new concept. Each of the parties was to cover its own expenses.

Conclusion

Substantial elements of innovation have been observable during the boom years in rig supply and organisation on the NCS. This trend has been driven partly by the fact that rising costs over many years had put profitability under pressure, and partly by the entry of new oil companies on the NCS with different needs from the large established players.

Innovation occurred along several dimensions. One is technical innovation, with the development of new and more specialised rig types. The idea is that units should be more cost-effective and productive when they are purpose-designed for more specialist tasks. New rig categories encountered resistance from contractors. The role for purpose-designed rigs is uncertain in a future Norwegian rig market. It will be easier to launch specialised rigs when the market is less tight. On the other hand, it may be easier to opt for a standardised rig at low rates. Much innovation has also occurred with contracts and organisation. Examples include changes to risk-sharing in contracts and vertical integration.

Some of these changes might be determined to a certain extent by economic conditions. One example could be oil companies owning rigs. Other innovations will represent lasting adjustments to collaborative relations between oil companies and contractors. New oil companies are less keen than the established players to build up a large internal staff to supervise drilling operations. This will mean a trend towards contractors taking on more functions than has been usual on the NCS. A greater use of turnkey contracts and integration of services can be seen. That makes big demands on the breadth of contractor knowledge, and calls for greater willingness and ability to bear risk. This is likely to create a number of challenges in a transitional phase.

References

Corts, K. (2000). *Turnkey Contracts as a Response to Incentive Problems: Evidence from the Offshore Drilling Industry*. Working paper, Harvard University.

Corts, K. S. and Singh, J. (2004). The Effect of Repeated Interaction on Contract Choice: Evidence from Offshore Drilling. *Journal of Law, Economics, and Organization*, 20(1), 2004, 230–260.

Hart, O. (1995). *Firms, Contract, and Financial Structure*. Oxford: Oxford University Press.

Hillier, B. (1997). *The Economics of Asymmetric Information*. London, Macmillan Press.

Milgrom, P. and Roberts, J. (1992). *Economics, Organization, and Management*. Englewood Cliffs, NJ, Prentice Hall.

Moomjian, C. A. (1999). Contractual insurance and risk allocation in the offshore drilling industry. *Drilling Contractor*, January/February, 19–21.

Osmundsen, P. (2013). Choice of development concept – Platform or subsea solution? Implications for the Recovery Factor. Oil & Gas Facilities (SPE), October, 64–70.

Osmundsen, P. (2010). Chasing reserves – Incentives and ownership. Bjørndal, E., Bjørndal, M., Pardalos, P. M., and Rönnqvist, M. (eds.), *Energy, Natural Resource and Environmental Economics*. Springer: Verlag Berlin Heidelberg, 19–39.

Osmundsen, P., Asche, F., Misund, B. and Mohn, K. (2006). Valuation of international oil companies. *Energy Journal*, 27(3), 49–64.

Osmundsen, P., Aven, T. and Vinnem, J. E. (2008). Safety, economic incentives and insurance. *Reliability Engineering & System Safety* 93(1), 137–143.

Osmundsen, P., Mohn, K., Asche, F. and Misund, B. (2007). Is the oil supply choked by financial markets? *Energy Policy*, 35(1), 467–474.

Osmundsen, P., Roll, K., and Tveterås, R. (2010). Exploration drilling productivity on the Norwegian shelf. *Journal of Petroleum Science and Engineering* 73, 122–128.

Osmundsen, P., Roll, K. H. and Tveterås, R. (2012). Drilling speed – the relevance of experience. *Energy Economics* 34, 786–794.

Osmundsen, P., Skjerpen, T. and Rosendahl, K. E. (2015). Understanding Rig Rate Formation in the Gulf of Mexico. *Energy Economics* 49, 430–439.

Osmundsen, P., Sørenes, T. and Toft, A. (2008). Drilling contracts and incentives. *Energy Policy*, 36(8), 3138–3144.

Osmundsen, P., Sørenes, T. and Toft, A. (2010). Offshore oil service contracts – New incentive schemes to promote drilling efficiency. *Journal of Petroleum Science and Engineering* 72, 220–228.

Osmundsen, P., Toft, A., and Dragvik, K. A. (2006). Design of drilling contracts – economic incentives and safety issues. *Energy Policy* 34, 2324–2329.

Additional sources

DN.no. (17.10.2007). Røkke bygger seg opp i Norskehavet [Røkke extends his presence in the Norwegian Sea]. *Dagens Næringsliv*. Retrieved 21.02.2018 www.dn.no/nyheter/energi/2007/10/17/rokke-bygger-opp-i-norskehavet.

DN.no. (25.08.2009). Røkke sluker Det Norske [Røkke devours Det Norske]. *Dagens Næringsliv*. Retrieved 21.02.2018 https://e24.no/makro-og-politikk/aker-bp/det-norske-og-aker-exploration-fusjonerer/3232536.

DN.no. (26.08.2013). Songa med store forsinkelser [Great delays for Songa]. *Dagens Næringsliv*. Retrieved 21.02.2018 www.dn.no/nyheter/energi/2013/08/26/songa-med-store-forsinkelser.

DN.no. (25.11.2013). Spleiselag berger Songa Offshore [Joint financing saves Sponga Offshore]. *Dagens Næringsliv*. Retrieved 21.02.2018 www.dn.no/nyheter/2013/11/25/spleiselag-berger-songa-offshore.

Improved Recovery Commission, 2010. (22.09.2010). Økt utvinning på norsk Kontinentalsokkel [Increased extraction on Norwegian continental shelf]. Report from an Expert Commission appointed by the Ministry of Petroleum and Energy, chaired by Knut Åm. Retrieved 22.02.2018 www.regjeringen.no/nb/dep/oed/dok/rapporter/2010/Okt-utvinning-pa-norsk-kontinentalsokkel.html?id=615841.

Rig Commission. (16.08.2012). Økt bore- og brønnaktivitet på norsk sokkel [Increased drilling and well activity on Norwegian shelfs]. Report from an Expert Commission appointed by the Ministry of Petroleum and Energy, 19.12.2011. Retrieved 22.02.2018 www.regjeringen.no/upload/OED/pdf%20filer/bore_og_br_aktivitet_riggutvalget_2012.pdf.

Part II

7 Born national – going global

Helge Ryggvik and Ole Andreas Engen

Introduction

This chapter discusses the internationalisation of the Norwegian offshore supply industry. Between the years 2000 and 2011 this industry's international turnover grew from 30 billion NOK to 196 billion – more than a six-fold increase. The offshore supply industry had become Norway's second largest export industry after the export of petroleum in the form of crude oil and gas, and before fish and aluminium that previously had been higher on the list. For the first time, advanced industrial products and services, not raw materials, dominated Norwegian exports.

Several aspects of the innovation policy that have been important in laying the foundations for the development of the Norwegian industry's competence are discussed elsewhere in this book (see Chapters 2, 3 and 14). This chapter focuses mainly on those conditions under which Norwegian companies have developed their international ambitions, continuous pathways towards other petroleum regions. Due to the many firms involved and continual changes due to mergers, acquisitions and break ups, it is impossible to present the full picture within the framework of a short chapter. With the intensified internationalisation from the early 1990s it becomes increasingly difficult to define what can be considered a Norwegian firm. In order to cover the long historical trends in this chapter we divide the internationalisation of the Norwegian supplier industry into three phases:

1 Early international expansion and protectionism 1970–1990:
The period from the discovery of the Ekofisk field until after the effects of the oil crisis was felt. The period also characterised by infant industrial policy through local content, goodwill agreements and incentives embedded in the concession practices.
2 High international ambitions 1990–2000:
The period when a number of the biggest Norwegian players declared internationalisation as a major goal, but where many were hit by huge losses when oil prices fell at the end of the 1990s. Implementation of the NORSOK programme.

3 International breakthrough 2000–2015:
The period where a wide range of Norwegian supplier industries ended up fully-fledged participants in global supply chains.

The chapter seeks to give a picture of factors that explain the development of a strong domestic supply industry, revealing the factors that shaped the technological competence and gave incentives to build up specialised clusters, and finally, enable Norwegian supplier companies to compete in a global arena and in some cases, to establish themselves abroad. The chapter is based on Ryggvik (2013) and on prime sources, i.e. interviews and documents, gathered and systemised during 2014–2015.

Early international expansion and protectionism

When the magnitude of the Ekofisk field became known in 1970, a large number of Norwegian shipping companies immediately threw themselves into the market for semi-submersible drilling rigs. With experience in international shipping since the eighteenth century, it was just not natural for Norwegian ship owners to restrict themselves to the Norwegian continental shelf. Soon, Norwegian ship-owners had the capacity to dominate the rig market on both the Norwegian and British continental shelves.

The network between Norwegian ship-owners and the shipbuilding industry was decisive when a new fleet of semi-submersible drilling platforms was built in Norway. Learning from its work on a rig built on licence for the American drilling company Odeco in the 1960s, the shipbuilding group Aker constructed its own semi-submersible rig concept. The first Aker H3 rig was built at a newly established shipyard in Verdal, inside the Trondheim fjord. In February 1974, Aker had orders for constructing 25 rigs, 11 on licence from other groups, several in other countries. Aker's semi-submersibles had become an early Norwegian export product.

The first production installations on the Ekofisk field, which constituted the largest volume and most technologically advanced part of the new offshore market, were produced abroad. However, the contract for the construction of a concrete tank for storing gas went to companies with a background in the Norwegian hydropower sector. Contrary to the shipping industry, this part of Norwegian industry had had a strong local, Norwegian orientation. Nevertheless, based on the engineers' first encounter with the petroleum industry, the newly established company Norwegian Contractors (NC) developed a concept for large production platforms with concrete under-carriages. The first major contracts applying this innovation did not go to sites on the Norwegian continental shelf, but to the British continental shelf. From 1973 until the last delivery in 1976, NC built four giant concrete substations that were exported to Britain (Hanisch and Nerheim, 1992).

The very first phase in the development of oil operations in the North Sea shows that both Norwegian capital and industry were well-equipped to find a place not only as suppliers for the expanding Norwegian oil industry, but also in establishing a place in the international production chain that dominated at the time. By the middle of the 1970s, the North Sea was overall the largest offshore market in the world. However, although the construction and control of a fleet of semi-submersible rigs and construction of the platform legs in concrete was important enough, both were some of the technologically least advanced parts of the production chain related to oil operations offshore. Oil companies sometimes used the term 'metal bashing' as a description of the work that had to be performed for the construction of platform legs and simple topside structures. Although shipping companies owned semi-submersible drilling rigs and drove the maritime part of the business, drilling equipment on board was operated by US drilling contractors. Similarly, when a steel jacket was completed by Norwegian yards, all drilling equipment was imported, mostly from American suppliers. Although platform legs in concrete called for major engineering calculations, the actual construction process was often carried out by workers with a wheelbarrow.

Protectionism

What from the early 1970s seemed to become an internationally integrated open offshore supply market, from the mid-1970s became more or less closed national markets. As early as 1971, with the so-called 'ten oil commandments', formulated by the Norwegian parliament, the development of a local supply industry was established as a central part of Norwegian oil policy (Innst. S., 1970–1971). A political goal to develop a local industry did not in itself imply a strict protectionist policy. However, concerns from the two largest shipbuilding groups, Aker and Kværner, were important when in 1972 the government stated in a royal decree: 'In cases where Norwegian commodities and services are competitive, in quality, service, delivery time and price, these are to be used' (Royal Decree, 1972). Despite the Ministry of Industry's insistence that '§54', as it was later called, was not a protectionist measure, most parties involved knew that if the government wanted to use the paragraph actively, this offered an opening for the government to force operators to increase Norwegian participation (NOU, 1979). Norwegian shipowners saw the paragraph as an immediate threat, both to their international shipping activities and to their recent successful breakthrough in the semi-submersible drilling rig market. The paragraph, however, had little significance before the international economic crisis following the sharp rise in oil prices struck in 1974. At this time, Britain was also taking measures to protect its oil industry.

It is impossible to estimate how the development of the Norwegian offshore supply industry would have been without the protectionist turn after 1974. Shipping companies which owned and operated the semi-submersible drilling rigs might eventually have evolved to become competent drilling contractors, irrespective of the protectionist regime. Moreover, many of the measures the Norwegian authorities supported would easily occur in trade regimes without open protectionism, like the strong efforts to adapt the education system and research institutions to the needs of the new industry. Other measures were in a border zone, such as when foreign oil companies had to support Norwegian research institutions as part of their 'Norwegianization obligations' in the work agreement in new concessions. However, in the following years when Statoil intervened to make sure that Norwegian firms acquired strategic contracts and were accepted as partners in joint ventures with experienced foreign firms, it was clearly a protectionist measure and essential for the development of firms that later were internationalised.

Deregulation in the late 1980s

While the introduction of the regime was clearly linked to the crisis in 1974, the factors leading to the break-up of the regime are more complex. Although Norwegian suppliers benefited from being located within the Norwegian side of the protectionist divide of the North Sea offshore supply markets, there was some frustration about being shut out of the UK offshore market, especially among the largest suppliers. In 1981, a public committee on the need for the internationalization of Norwegian business referred to the trade barriers after the oil crisis as 'new-protectionism' and concluded the best strategy for Norwegian companies was to circumvent these barriers by purchasing well-established local companies (NOU, 1981). Foreign direct investment was at this time first and foremost seen as a means of sustaining exports. From the mid-1970s and in the 1980s, Norwegian foreign direct investments in general increased slightly, despite the form of new-protectionism in the petroleum sector (Amdam, 2009). In the oil sector, some of the largest Norwegian suppliers at the time, Smedvig (the largest rig operator), Aker and Kværner (engineering and construction) actually tried to circumvent British protectionist measures through acquisitions. However, none of these acquisitions abroad in the 1980s was a big success.

The 'new protectionism' in the North Sea offshore supply markets appear as a paradox since the late 1970s and the early 1980s many countries were characterized by a revival of liberalism. However, with massive growth in foreign investment from the second half of the 1980s it was clear that a more integrated interwoven global economy was also influencing the international oil industry. Dramatically falling oil prices in the second half of the 1980s accelerated the international orientation. Statoil, which at this time had become the largest operator on the Norwegian continental shelf, wanted increased degrees of freedom choosing the cheapest and best supplier, regardless of the company's national affiliation. At the same time, a common understanding within the industry evolved assuming that many Norwegian suppliers had developed sufficient skills to meet international competition. The 'infant industry policy' seemed to have been a success. The decisive political change, however, was the launch of the EU internal market as an idea in 1985. It was clear that protectionism in the oil sector was not in accordance with the free movement of goods, services, people and capital that the European economy was to be based upon. The single market was not formally launched before January 1993. Norway was not part of the EU internal market until the implementation of the EEA agreement in 1994. However, as negotiations started in the late 1980s, both Norway and Britain liquidated the openly protectionist measures.

High international ambitions 1990–2000

The Norwegian oil industry's international ambitions became publicly visible when, in August 1990, Statoil announced that the company had entered a strategic alliance with the oil company BP (Ryggvik, 2002). The two companies were to operate jointly in several countries which were opened up for international investment following the break-up of the Soviet Union and its allies. As a state-owned company in a preferred position in all large oil fields, Statoil was rich in capital and therefore in a much better financial position than the Norwegian companies in the supply sector. When Norwegian politicians enthusiastically supported Statoil's ambitious but risky move, a central argument was that Statoil's international activities would be helpful for Norwegian suppliers (White Paper 26, 1993–1994).

However, when Statoil and also the other two Norwegian oil companies, Norsk Hydro and Saga, launched their new international ambitions, several of the largest Norwegian suppliers were already on their way at full speed out in the world. In fact, despite high ambitions, it took decades before Statoil acquired a substantial operatorship abroad and by that was in a position to hire contractors. Hence, when several Norwegian supply firms experienced an international breakthrough, long before the Norwegian oil companies, their relations with international oil companies operating in Norway were more important than relations with Statoil.

The first years of the 1990s became a turning point in the sense that international companies acquired a number of newly-established Norwegian offshore suppliers. On a larger scale, companies such as German Siemens, the Swedish-Swiss ABB, and the American FMC, acquired Norwegian firms so as to gain access to what at the time was a growing market and considered as a leading region in the development of subsea technology. Turnover in the Norwegian oil sector grew from around 30 billion NOK in 1990 and reached a peak of nearly 80 billion NOK in 1998, despite stagnant oil prices. In 1996, a Norwegian industry association, the Confederation of Norwegian Industry (NHO), estimated the market for offshore petroleum activities worldwide to be 201 billion NOK, with the Norwegian part at the time corresponding to a quarter of the total.

The opening of international competition related to the Norwegian offshore supply industry from the early 1990s led almost immediately to radical changes in the industry's structure. From 1990, there was an almost continuous wave of mergers and acquisitions, but also of splitting and reorganizing corporate units. The same geographical unit of manufacturing halls, production equipment and workers could experience changing names every other year.

The Norwegian suppliers' international turnover reached a temporary peak in 1995 of around 18 billion NOK (INTSOK, 1996). Although several of the largest companies had a clear global orientation, it was contracts associated with the British continental shelf that helped keep the turnover up (9.5 billion NOK) (Heum *et al.*, 2006). Revenue related to the UK market was almost equally divided between direct exports from Norway and sales through subsidiaries. The relationship between export and sales through subsidiaries was clearly different between the different segments of the supply industry. In areas such as seismic and reservoir analysis, electro/instruments and marine services, most of the international turnover was in the form of exports. In the mechanical equipment sector, which had the largest turnover (4.4 billion NOK), half of the turnover was through subsidiaries. In the drilling and well service industries, engineering, operation and maintenance services, more than 70 per cent of sales were conducted through subsidiaries.

Unsuccessful investment in construction

Aker's main strategic goal in this period was to get into a position to acquire major engineering, procurement and construction (EPC) contracts in the construction of hulls as well as topsides. When Aker, now under the name Aker Maritime, acquired the Newcastle McNulty offshore yard in 1996, it hoped to be in position to do that in the UK sector. The Newcastle division attempted to specialize in floating production concepts or so-called TLPs (Tension-leg platforms) that were particularly suitable for relatively deep parts of the North Sea. However, despite the renewed capacity, Aker struggled to capture a large contract on the British continental shelf. The US offshore market was equally challenging for Aker. In 1991, the company acquired the Houston-based engineering firm Omega Marine. Immediately after, Aker secured a 51 per cent stake in the offshore yard Gulf Marine Fabricators (Aker, 1993, 43). This yard had constructed the world's two largest sub-structures in steel. The company had about 600 employees when Aker took over, and had the capacity to expand. By that, Aker hoped to have the capacity to take on the role of an EPC supplier of large production installations in the Gulf of Mexico. However, buying a majority stake in Gulf Marine Fabricators was bad timing. Hereafter, all new large deep-sea installations used more complex floating installations.

Kværner carried out a corresponding but even more ambitious round of international acquisitions in the 1990s. Following acquisitions in Great Britain, Kværner also established a solid position in Asia. Kværner bought up two significant engineering companies which were merged (Humphreys and Glasgow Engineering, and Earl and Wright). But similar to Aker, Kværner's main problem was also to establish itself as an EPC contractor for the largest construction projects directed by the big oil companies. In the UK, as on the Norwegian shelf, the largest companies had picked a few large contractors, relying on their ability to deliver both standardised technical solutions and the necessary innovation while at the same time ensuring the certain competition.

In many ways, Aker and Kværner were hit by the same kind of contractual politics that had contributed to securing their position in the Norwegian sector where they were both established EPC suppliers to large projects. A remaining alternative was thus to acquire one of the established British offshore contractors. In 1995, Kværner attempted to buy up the large British construction company Amec. However, the attempted acquisition met with so much opposition that Kværner decided to withdraw. A few months later, in January 1996, Kværner bought the major British company Trafalgar House. With the company's 23,000 employees, Kværner had become a giant. But Trafalgar House was a conglomerate whose offshore component was only one of many other large sections. Kværner's leader, Erik Tønnseth, was accused of suffering from megalomania.

The critics were proved right. The attempt to become a global player in a period with low oil prices broke Kværner. Furthermore, at the end of the 1990s there was little to be had from the Norwegian continental shelf. At the end of the 1990s and early 2000s, there was both a major restructuring and a transformation of ownership on the construction side of the Norwegian industry. Aker's owner from 1995, Kjell Inge Røkke, managed to gain control of what was left of Kværner. In 2004, Aker and Kværner were in practice merged into a single offshore giant.

In the meantime, Aker had established a promising position in the US construction market for hulls for large deep-sea installations. In 1995, Aker bought the US company Deep Oil Technology. This was originally a Finnish-American company that had developed the so-called 'Spar concept', a floating tubular structure, suitable for carrying a large production platform and with capacity for permanent deep-water production drilling. The actual construction was carried out by a Finnish company, Rauma Offshore, which Aker also took over. The acquisition of Rauma Offshore also proved to be a bridgehead into the offshore industry in Kazakhstan and Azerbaijan. The Finnish company had relations with the former Soviet republics from the period prior to the collapse of the Soviet Union.

Due to financial constraints, around the year 2000 most of Aker's and Kværner's international acquisitions on the construction side were now sold. This also included the acquisition of a yard in Brazil (Chapter 8). With its dominance in Norway, by acquiring Kværner, Aker could count on getting a large part of new field development projects in Norway in the 2000s. However, given new technologies and fewer large oil finds, the kind of mega-projects that had characterized many fields' development projects up until the 1990s were less frequent. At Aker's Norwegian yards, an increasing share of the workforce was moved from traditional construction into a growing market for maintenance and modification of a growing number of installations. In fact, these were activities that Aker had learned from some of its earlier and successful acquisitions in Britain (Aker, 1998). However, when the offshore contractor market increased again, Aker's main focus was on the subsea part of the industry.

Rig operators

On the exploration side of the industry, the internationalization of Norwegian actors followed a pattern similar to that of construction in the sense that the first part of the 1990s was characterized by expansion while the second part was characterized by problems and withdrawals. The market for offshore rigs had suffered overcapacity since the late 1970s. The protectionist barriers between the Norwegian and British markets did not improve the situation. Following a number of mergers and acquisitions at the beginning of 1990, Aker Drilling was the largest drilling operator, with Smedvig as a good number two. For both companies, the year 1990 marked a turning point with large acquisitions of UK registered companies. Aker Drilling bought the British Transocean Drilling Company Ltd., while family-owned Smedvig bought the British-based company Robray Offshore Drilling. Robray had a significant rig fleet in Asia.

The main motive for the acquisitions was, as on the construction side, to obtain market access. However, while construction continued to be characterized by informal ties to local suppliers, the impact of deregulation that followed the EU internal market was more immediate for the rig market. Since there was overcapacity on the international rig market, acquisitions were often followed up with the upgrading of some of the oldest rigs. Aker Drilling, which shortly after the acquisition in the UK changed its name to Transocean, expanded until 1995 to become one of the world's largest rig operators. About half of the turnover occurred in operations outside the Norwegian continental shelf. For Smedvig, the acquisition of Robray meant more than half of the company's turnover originated abroad.

The combination of mergers and acquisitions and the reduction of capacity had the desired effect that the rig-rates and hence profitability went up. However, Transocean, a joint stock company, was particularly vulnerable to acquisitions. In 1996, Transocean was acquired by the American drilling company Sonat Offshore Drilling in a hostile takeover. During the sale, drilling from fixed Norwegian production platforms was separated out. Sonat, which now took over the name Transocean, retained a position in the exploration rig market on the Norwegian continental shelf. However, the new Transocean's 'Norwegianness' was heavily diluted when the company commenced operations abroad and continuing into the 2000s attained the role of the world's largest drilling company offshore (see also Chapter 6).

International breakthrough 2000–2015

Whereas Aker and Kværner had to pull out of international construction projects, other parts of the Norwegian offshore supply industry experienced a more continuous growth. From 2002, the industry's total turnover increased again. From 2004, growth was exceptionally large. Between 2000 and 2011, the industry's international turnover grew from 30 billion NOK to 152 billion: a fivefold increase! The Norwegian offshore supply and service industry had made its decisive breakthrough as a competitive global technological contractor industry. It was this growth that made it Norway's second-largest export business after sales of oil and ahead of fish and aluminium which had previously been higher on the list. This breakthrough was also important in a historical perspective: for the first time in Norway's history, it was not the sale of raw materials alone, but advanced services and equipment which dominated the list of export products.

Subsea installations

The successful growth in the Norwegian offshore industry after 2000 has more than anything else been associated with subsea technology. During the first half of the 1990s, three companies were established as major suppliers of subsea structures for the Norwegian shelf: Kongsberg Offshore, Kværner and ABB. But even if the three companies produced very similar products for a period, their routes into the growing subsea market were different. There were also large differences in the degree of 'Norwegianness' when it came to ownership, affiliation to the Norwegian engineering communities, and the approval and inclusion in what eventually became known as global production chains.

The largest supplier of subsea constructions on the NCS in the 1990s was Kongsberg Offshore. Originally, Kongsberg was part of the state-owned company Kongsberg Weapons Factory (KV). When the company was divorced from KV in 1987, Siemens first took over as owner. As part of the protectionist regime, KV had secured a contract for delivery of christmas trees[1] for several Norwegian fields. The deliveries were initially manufactured under licence with most key parts imported. KV had in the first years a licence agreement with the American company Cooper Cameron. Later, the company entered into a similar agreement with the company Food, Machinery, Chemicals (FMC).

The Norwegian engineering company at Kongsberg soon developed its own technology. The breakthrough came when the company received a very large development contract for standardized subsea installations for Statoil. When FMC took over as an owner in 1993, it was not only to ensure access to the Norwegian market, but to also acquire the company's expertise. This was expressed by the fact that the Kongsberg-part of the FMC from the late 1990s was responsible for exporting underwater installations in the FMC Group internationally. Kongsberg Offshore's experience in the 1990s was confirmation of an emphasized topic in today's globalization literature, emphasizing that competence, not ownership, was crucial for local value creation.

ABB secured a similar status as EPC supplier of subsea technology with a contract for the delivery of equipment to the private Norwegian oil company Saga. New to the offshore scene, ABB achieved this position as a result of a number of acquisitions of both Norwegian and international suppliers. The ABB group located the company's new global offshore division to Norway and several international procurements. US Vecto Gray and the British Global Engineering secured key units for the development of new technology. This turned out to be crucial when ownership changed. When ABB pulled out of the offshore subsea segment in 2003, segregated pieces of the company continued under the name of Vecto Gray, a company that soon emerged as an international subsea supplier without a strong Norwegian association.

The third route into the global subsea market was via the Norwegian Kværner company. Although Kværner competed with Aker as a supplier of large offshore installations in Norway, the company distinguished itself from Aker in the sense that it traditionally had been a supplier of various types of advanced mechanical equipment. The company delivered subsea units based on its own designs already in the 1980s. In the 1990s, Kværner received a large contract for the supply of underwater equipment for the second largest Norwegian oil company, Norsk Hydro. Based on the current experience, Kværner managed to secure a substantial contract for the delivery of christmas trees associated with the development of dive-free installations located in deep water in Brazil in the late 1990s.

When Aker took over Kværner in the early 2000s following unsuccessful attempts to buy several large ailing shipyards, Kværner's far more successful subsea division would be at the heart of the company's further development in the 2000s. The company operated under the name of Aker-Kværner for a period. Since 2008, the company's name has been Aker Solutions. In 2008, Aker Solutions had 23,360 employees, more than 50 per cent of whom were employed outside Norway. In subsequent years there was a consolidation and focus, resulting in the company operating under two main divisions from 2015: one for subsea equipment and another for field design and maintenance, modification, and operations. In 2011, the traditional construction division that had undertaken large construction assignments on the Norwegian shelf became a new company under the original Kværner-name, and in 2013, the Department for Well Intervention was also sold. By 2015, a total of 7,673 (44 per cent) of Aker Solution's employees worked in Norway. Of the company total of 17,673 employees, 2,048 worked in the United Kingdom, 2,994 worked in Asia, 1,370 in Brazil, and 868 in North America.

Even though offshore services to the Norwegian market were important to Aker, the company and especially the subsea division, appeared as globally integrated. In 2000, Aker invested in three new factories for the production of advanced underwater equipment in Norway (Tranby), Singapore and Brazil (Curitiba). Control systems for the same installations were delivered from a factory in Aberdeen, while advanced umbilicals were delivered from a specialized factory in Malaysia. On the other hand, they sold the factory in the US where Aker (like Kværner) first acquired the necessary construction expertise. Although Aker Solutions retained a strong Norwegian anchorage, the company developed a globalized production chain. Aker's Norwegian division was a significant exporter and conducted most of the international sales through subsidiaries. At the same time, there was much trade between the different units. This implied that Aker also appeared as a significant importer.

Subsea-services

The part of the subsea segment that took on services such as installation, inspection and modification of underwater equipment grew largely out of the same part of the supply industry that in the 1970s and 1980s undertook advanced diving. The technological development experienced many leaps (Gjerde and Ryggvik, 2013). In 1992, Stolt-Nielsen's diving company, Stolt-Nielsen Seaway, was merged with the well-experienced French diving firm Comex to form the Stolt Comex Seaway company. This was a real Norwegian takeover of ownership. However, the once-French Comex had gradually become more British. The headquarters for the new company were moved to Aberdeen. In the years which followed, the company broadened its international footfall, making significant purchases both in the US and France. From 2003 the name was changed to Acergy.

The company Subsea 7 was the result of a complicated process of buyouts and mergers of companies with a position in the Norwegian subsea market made by the Norwegian shipowner Kristian Siem. In 2002, Subsea 7 acquired the diving-related part of Halliburton, which for its part had taken over the American Taylor Diving, a key player in advanced diving in the 1970s. With his large share, Siem was the chairman of the board. However, his Norwegian affiliation was limited. The company's shares were traded on the Oslo Stock Exchange, it is true, but like Stolt-Nilsen, Siem transferred his address of residence from Norway. Subsea 7's top leadership comprised senior managers with backgrounds in American firms. It would be necessary to go to the firm's local department in Stavanger before one found many Norwegians in the organizational hierarchy.

In the early 2000s, Subsea 7 competed in the same market as British-Norwegian Acergy and the American Oceaneering. Finally, in June 2010, Acergy acquired Subsea 7. Siem kept an ownership share of 20 per cent and remained chairman of the board. In 2014, the company was unchallenged as the world's largest subsea engineering or seabed to surface firm. In 2014, just before the downturn, the total workforce was 13,000 (Subsea 7, 2014). The company operated 39 specialized vessels, with five more under construction. The company was traded on the Oslo Stock Exchange, but its operations headquarters were in London.

The second revival of Norwegian drilling

In general, and after losing control of Transocean and experiencing a difficult time for Norwegian offshore drilling companies lasting until the early 2000s, the Norwegian drilling industry subsequently returned in full force. The family-owned firm Odfjell Drilling, which in the 1990s had survived by being cautious and concentrating on the Norwegian market, now acquired new rigs with the capacity to drill in deep waters, and secured contracts in Tanzania and Brazil. A new Røkke-controlled company under the old name Aker Drilling re-emerged, with brand new rigs.

The greatest change came in 2005, when the shipowner John Fredriksen, through a series of acquisitions of Norwegian firms, including Smedvig, established the firm Seadrill. Fredriksen was the owner of the Bermuda-registered Frontline, the world's largest tanker shipping firm. Seadrill started as a Norwegian limited company with its headquarters in Stavanger. Within a few years, following a combination of mergers and acquisitions all over the world and ordering new rigs, the company established a fleet that covered the entire global offshore market. If installations on order are included, in 2011 Seadrill had nine large drill-ships, 14 semi-submersible rigs, 21 jack-ups and 22 tender rigs of various kinds. The company was now competing seriously with Transocean to become the largest in the global offshore drilling market.

Once again, the question is 'how Norwegian was Seadrill really?' Even though Seadrill was registered as a Norwegian limited company and was traded on the Oslo Stock Exchange, its many rigs were registered in Bermuda and under other flags of convenience. With its headquarters in Stavanger the company in itself accounted for a significant part of international turnover statistics. So when the company decided to move its headquarters out of Norway in late 2012, its 35 billion NOK international turnover was no longer included in the statistics of Norwegian companies' international turnover (Rystad, 2016). This contributed to the decline in the total Norwegian offshore-related turnover from 183 billion NOK in 2012, to 166 billion the following year, even though the general trend for Norwegian global presence was still growing.

Maritime cluster for international growth

The oil boom in the early 2000s also represented an international breakthrough for Norwegian offshore supply vessels. With the exception of Siem Offshore, which was established when a number of ships were sold by Subsea 7, the companies that dominated this part of the Norwegian offshore segment grew out of companies that originally operated local Norwegian coastal transport. This meant that they traditionally had a more local, Norwegian orientation than major shipping companies. In return, the supply boat industry operated in a tight symbiosis with a series of smaller shipyards that specialised in the construction of small and medium-sized special vessels. Unlike larger, traditional shipyards, which had either been converted to offshore yards or shut down in the 1980s, this part of the Norwegian shipbuilding industry survived. Nevertheless, this did not occur without major changes. In order to reduce costs, several of the yards chose to manufacture hulls in Eastern Europe. However, all engineering work and installation of equipment was carried out at the shipyards in an area of north-western Norway which, together with the owners and operators of vessels that were built in the 1990s, was designated as a Norwegian offshore-related maritime cluster.

It is reasonable to assume that the traditional coastal Norwegian affiliations were a contributing factor in a small number of supply vessel owners following other parts of the supply industry into the world from the early 1990s. During the oil boom in the 2000s, in contrast, the international expansion was extensive. Companies such as DOF ASA, Farstad Shipping, Solstad Offshore, and Eidesvik Shipping ordered a number of new ships to participate in international missions. In addition, companies such as Siem Offshore and the newly established Fredriksen-controlled Deep Sea Supply entered the same market.

The demand for offshore supply vessels internationally also reflected the overall growth in the offshore market. From the early 2000s to 2012, the international supply fleet experienced a threefold increase. Growth took place in new markets such as Brazil, West Africa, Asia and Australia. Norwegian companies secured a considerable market share in all these markets. Companies that had operated primarily on the Norwegian continental shelf in the early 2000s became global companies ten years later with market shares distributed throughout all central offshore regions.

The extensive expansion, however, appeared to lead to a crisis, regardless of the fall in oil prices from 2014. An international market analysis by Farstad Shipping from 2012 pointed out that the great expansion had led to overcapacity in the market. Most major players had orders for a number of new ships. Accordingly, when the crisis struck the hardest, those companies that had expanded heavily were hit hardest. The freight rates showed a declining trend even before the oil price fall in 2014. The combination of cases revenue and rates, which in some cases made it impossible to operate without loss, pushed a number of players towards bankruptcy. The time was nigh for a comprehensive restructuring of the industry.

The drilling package cluster

The subsector that experienced the most remarkable export growth in the 2000s was drilling equipment, pumps and hydraulic cranes specially adapted for semi-submersible drilling rigs and drill-ships. Cranes and pumps also had a market in an increasing number of production ships (FPSOs). When drilling in the North Sea commenced, there was already a large American industry specialized in various types of drilling equipment. Norwegian companies discovered that each individual offshore drilling operation required equipment to be specially adapted to the ocean depth, drilling depth and the particular geological formations. Moreover, towards the end of the 1980s the drilling industry went through various important technological changes. Much of the dangerous work on the production deck was replaced by automated equipment. This was increasingly steered from control rooms using IT and sensors. At the same time, there was a greater use of horizontal drilling, which posed even greater demands on the control systems.

It was a group of companies based in the Kristiansand area (on the southern tip of Norway) with a background in traditional ship equipment that managed to break into the parallel, but more advanced market for key equipment on drilling rigs at sea. In the early 1990s, the engineer Bjarne Skeie established a company that specialized in hydraulics and the development of specialised cranes for oil installations. Because of the danger of explosion in electrical equipment, hydraulics were an important tool for various lifting processes in drilling. Once again, strict Norwegian safety requirements followed by increased automation, together with a Norwegian company with a high proportion of engineers, proved a winning combination. By the end of the 1990s, Hydralift had become a significant exporter of specialised equipment.

In 2000, Hydralift took a further step towards being able to deliver complete drilling packages by buying up the engineering division of Procon (a part of Transocean that had remained in Norway after the American acquisition in 1996). With Procon's engineers, the company had secured part of Norway's core drilling expertise. Transocean could take rigs and equipment with them and acquire control over foreign assets, but the actual Norwegian skill in the form of Norwegian engineers and workers was not easy to move. In 2002, Hydralift was in turn bought up by the large Houston-based American drilling equipment supplier National Oilwell. National Oilwell was really one of Hydralift's competitors. However, it turned out that the Norwegian division was given great freedom to develop within National Oilwell Varco (NOV). By 2012, the company had an export share of 80 per cent. It was one of Norway's five largest exporters.

In 2005, the companies regarded as belonging to the cluster in the Kristiansand area had 1,800 employees and a turnover of 2.5 billion NOK. In 2012, there were about 9,000 people working in the cluster, with a turnover of up to 46 billion NOK. What is most striking in this development is the transition from construction to an emphasis on advanced engineering skills. A majority of the employees had some form of engineering expertise. Only a small part of the physical equipment constituting the drilling packages supplied to shipowners, drilling companies and oil operators was built in Norway. Norwegian engineers were involved in developing and adapting the equipment, but the most important part of the work lies in being able to assemble the various pieces of equipment together into a packet which actually works. The knowledge base is centred on a sense of the whole and the ability to combine various specialisations within the engineering profession. The equipment in question had to be produced, delivered and maintained, but it also had to comply with the requirements of various industry standards (NORSOK, API, ISO and so forth), and the safety regulations of the countries in question. Typical buyers were Norwegian companies ordering FPSO, drill ships and semi-submersible rigs on these yards, most often linked to instructions that the equipment was to comply with the Norwegian NORSOK-standard.

However, the explosive growth from 2005 in the Kristiansand-cluster was strongly linked to the historical replacement and extension of drilling equipment which marked this phase. It was also an issue that so much of the expertise in the cluster was placed inside one foreign-owned company. When the crisis arose in 2014 there was a genuine fear that the American owner NOV would prefer employees in the US when costs had to be cut. A 2016 report from the consultancy firm Rystad showed that Norwegian offshore supply companies which were largely based on exports fared better in the crisis than companies which were mainly focussed on the Norwegian market (Rystad, 2016). However, as an exception to that trend, the Kristiansand-cluster was hit particularly hard. From early 2015 until mid-2016, NOV reduced the number of its employees in Norway from more than 5,000 to fewer than 2,000 (E24 2016). Many of these were working in the Kristiansand area.

Small and global

At the end of the 1990s, ten companies accounted for more than 60 per cent of Norwegian-based companies' international turnover. In 2015, according to Rystad Energy, 20 companies accounted for a total of 75 per cent of the global turnover. On the list of the 20 biggest companies, there were six companies with parent companies in other countries. Of other companies in the top-20 list that until now we have not discussed, BW Offshore had developed into a major owner and operator of FPSO vessels. Another sector where Norwegian units formed the basis for major international activity was seismic. Since its establishment in the early 1990s, the Norwegian-owned Petroleum Geo Services (PGS) has become a leading global player. On the list one can also find the French–American Geoservice company CGG, which accounted for significant exports on the basis of its Norwegian branch. Although Kværner Yards, which operated the former Aker shipyards in Verdal and Stord, was spun off from Aker with the main objective of focussing on the Norwegian market, its international activities were large enough to become one of Norway's largest exporters.

When a limited group of companies accounted for a large proportion of the Norwegian supplier industry's international turnover, one natural explanation was that significant costs associated with becoming established in foreign markets would be more burdensome for smaller players. Typical of the group that was outside the list of the top 20 Norwegian internationals was that a relatively larger share of international turnover took place in the form of exports directly from Norway (about 60 per cent) (Rystad, 2016). Nevertheless, with more than a six-fold increase in the global turnover from 2000 to 2014, the international turnover of companies which found themselves outside the top 20 around the millennium had now also become significant.

In the group of companies which were closest to the 20 largest, there were companies that did not materially differ from the companies that were on the list. For example, certain supply shipping companies had a global presence, but were not large enough to compete. Thereafter follow the large group of small and medium-sized subcontractors. This group consists of so many different types of companies, targeting almost all of the many sub-segments of the supply industry, that it is difficult to give any coherent picture of how they have established themselves internationally.

Some subcontractors have followed the same pattern as larger companies in that they also have seen the opportunity in foreign markets and by themselves have become engaged in tenders internationally in order to win contracts.

Others have created smaller subsidiaries, either by buying up smaller local firms or by building up new organizations from scratch. For small companies, the foreign subsidiaries will often be only sales offices, and thus a means of achieving increased exports. It appears, however, that smaller companies have often had greater benefit from Norwegian organizations such as Innovation Norway and INTSOK (from 2017 Norwegian Energy Partner (NEP)), whose aim is precisely that of supporting the smaller players.

Many smaller subcontractors were naturally drawn onto the world stage because of their relations to larger suppliers. Major Norwegian suppliers might prefer to employ Norwegian subcontractors they know well from previously instead of building up a base of local subcontractors in countries where they operated. In many parts of the industry, knowledge of the technical standards employed by the major companies was essential. In some cases, moreover, smaller Norwegian companies were drawn into the globalised offshore supply chain, not because they themselves had not taken an initiative, but because their competitors had established themselves abroad.

Throughout the entire period there were examples of Norwegian companies that did not succeed in the increasingly globalised offshore supply chain. In periods of low oil prices several companies collapsed or were sold cheaply as a result of a strong exposure to foreign competition. Nevertheless, the evolvement of the Norwegian offshore supply and service industry from the opening of international competition in the early 1990s has been a success, at least until the slump in oil prices after 2014. Even though several firms have been struggling since, there are indications that in spite of the fact that markets as a whole have been shrinking, the relative position of Norwegian companies has been increasing.

Note

1 A vertical assembly of mechanical elements used in oil exploration and production in surface and underwater oil and gas wells, primarily for flow control. Its name is derived from its shape, which roughly resembles a christmas tree.

References

Amdam, R.P. (2009). The internationalisation process theory and the internationalisation of Norwegian firms, 1945 to 1980. *Business History*, 51(3), 445–461.

Gjerde, K.Ø. and Ryggvik, H. (2013). *On the edge, Under water*. Stavanger, NO: Wigestrand Forlag.

Hanisch, T.J. and Nerheim, G. (1992). Fra vanntro til overmot? *Norsk Oljehistorie 1*. Oslo.Hanisch, T. and Nerheim, G. (1992). Norsk Oljehistorie. Fra vantro til overmot? [The History of Norwegian Oil: From Disbelief to Arrogance?] Oslo: Leseselskapet.

Heum, P., Vatne, E. and Kristiansen, F. (2006). *Petroleumsrettet næringsliv i Norge: Tiltakende internasjonalisering og global tilstedeværelse* [Petroleum-oriented business in Norway: Increasing Intrnationalization and global presence]. *SNF-Arbeidsnotat* Nr. 37/06.

List, F. (1841). *The National Systems of Political Economy*. London. Open Library of Liberty.

Ryggvik, H. (2013). *Building a skilled national offshore oil industry*. Oslo: Confederation of Norwegian Enterprise (NHO).
Ryggvik, H. (2002). *BP/Statoil alliansen. Et samarbeid til besvær* [An upsetting collaborative effort], TIK Working paper No. 16/2002. Oslo: TIK: University of Oslo.
Rystad Energy. (2016). *Internasjonal omsetning fra norske oljeserviceselskaper* [International revenue from Norwegian oil service companies]. Report to the Ministry of Petroleum and Energy 25. Oslo: Rystad Energy.

Additional sources

Aker, annual reports 1991–2000.
E24. (22.04.2016). National Oilwell 'Varco kutter 520 ansatte' ['Varco cuts 520 employees'] (Newpaper article). Retrieved 28.02.2018.
Innst. S. no. 294 (1970–1971) (Recommendation to Parliament). Retrieved 28.02.2018 www.stortinget.no/no/Saker-og-publikasjoner/Stortingsforhandlinger/Lesevisning/?p=1970-71&paid=6&wid=a&psid=DIVL2203&s=True.
INTSOK. (1996). En spørreundersøkelse om internasjonal virksomhet i norsk leverandørindustri. By Elling Homsø, Markeds-Forum Offshore for Intsok. [Survey about international activities].
NOU [Official Norwegian report] 1979: 43. Petroleumslov med forskrifter [Petroleum law and regulations].
NOU [Official Norwegian report] 1981: 47. Behovet for internasjonalisering av norsk næringsliv [The demand for internationalisation of the Norwegian Industry].
Royal Decree of December 8th 1972. Retrieved 28.02.2018 www.npd.no/en/Regulations/Regulations/Norm-price-fixing/.
Subsea 7, Annual Report 2014.
White Paper 26. (1993–1994). St. meld 26. (1993–1994), Utfordringer og perspektiver for petroleumssektoren [Challenges and Perspectives for the Petroleum Industry on the Continental Shelf].

8 Norwegian suppliers in Brazil

Helge Ryggvik, Ole Andreas Engen and Antonio José Junqueira Botelho

Introduction

This chapter analyses how Norwegian firms and Norwegian competence developed a large presence in the Brazilian offshore supply market. In a broader historical perspective, the Norwegian supply industry's significant presence in the Brazilian offshore market is interesting, not least because drilling for petroleum offshore commenced at the same time in Brazil and Norway. In both countries, there were early attempts to build a local industry related to the activities. The fact that many Norwegian companies now found themselves in a teacher role in many segments of the Brazilian oil industry was in itself a sign of the Norwegian industry success.

In 2015, exports to Brazil by Norwegian offshore supply and service companies amounted to 26 billion NOK (E24, 2016). Brazil's importance to the Norwegian oil industry was even more important than could be read out of the high export figures. A significant part of the Norwegian exports of specialised equipment, engineering services and other services to South Korea and other Asian countries were mounted on installations that ultimately were used in Brazil. The Brazilian market was also becoming very important for companies of Norwegian origin which had moved out of Norway, but which helped put a 'Norwegian stamp' on the Brazilian oil industry. This chapter is based on Ryggvik 2018, and on interview data gathered by Ryggvik, Botelho and Engen in Brazil in spring 2016.

The Brazilian oil experience

In order to understand how the Norwegian petroleum industry encountered Brazil, we have to briefly go through Brazilian petroleum history. As the search for oil offshore began in Brazil in the mid-1960s, in contrast to Norway, Brazil had already a long history of exploration for oil on land. With 49 per cent of the South American landmass and several large sedimentary basins, many hoped that Brazil, just like neighbours Venezuela and Mexico to the north, also sat on huge oil reserves. Like many other Latin American countries, however, there was also significant political resistance against the major foreign oil companies.

Therefore, when the state-owned Brazilian oil company Petrobrás was founded in 1953, it received a monopoly on all exploration and extraction of oil (de Oliveira, 2011; Priest 2016; Randall 1993). Petrobrás soon found several smaller fields on land. Nevertheless, not even the only significant discovery, the Carampolis field, discovered in 1962, was sufficient to ensure Brazil a breakthrough as a major oil producer.

However, Petrobrás geologists noted that both the Carampolis and other smaller discoveries were located in geological sedimentary structures that extended out into the Brazilian continental shelf. Many geologists and engineers in Petrobrás had their education from the US and were familiar with the development of technology for oil drilling offshore in the US Gulf of Mexico. American experts where hired to undertake the first offshore drilling in Brazilian waters too, and soon after, in 1968, the Garupa field was discovered.

The Garupa field was small compared with the series of major new discoveries that followed soon after in the North Sea. In the first part of the 1970s, in the absence of major new discoveries, the issue of increased foreign participation in the oil industry was raised anew. Hence, in 1975, former Petrobrás leader and Brazil's president Ernesto Geisel announced that large areas onshore and offshore should be opened up to foreign oil companies. Brazil was thus turning in the opposite direction of what was the main trend internationally. While both OPEC and other oil-producing countries, such as Norway, tightened the conditions for foreign companies, large areas in Brazil between 1976 and 1988 were open for foreign participation.

Nevertheless, the first significant growth in oil production offshore came in the areas of the Brazilian continental shelf where Petrobrás still dominated. From 1975, the year of the opening, Petrobrás made a series of new medium-sized discoveries in the shallow waters of the Campos Basin. The findings contributed to a marked growth in oil production from the early 1980s. The crucial breakthrough for Brazil as an oil producer, however, came in 1984 when Petrobrás first found the Baracuda field, and the Marlim field in the following year. Although still in the Campos Basin, both fields were discovered further out and in deeper water than previous findings. Baracuda was at about 300 metres depth while the Marlim was all of 800 metres. In the following years and into the 1990s Petrobrás made a series of new discoveries in ever greater depths in the same areas.

The new findings were in a borderland where it was possible to operate using contemporary technology. It therefore took many years before suitable production facilities were in place. However, the development that took place from the late 1990s led not only to Brazil approaching the level of production in Norway and Great Britain, it also represented a large market for the international offshore supply industry which had the technology Brazil now needed.

The findings in Campos, first in shallow and then in deeper waters, was seen as a success for Petrobrás. Since its foundation Petrobrás had focused on building up a large, competent group of geologists. In the early years most of these were foreigners. In the 1960s, after the capacity to train geologists in Brazil was increased, most geologists were Brazilians. With the exception of Shell which, through a Brazilian ally discovered a major gas field (Merluza), there were no other major discoveries in the many blocks that were opened to foreign oil companies between 1976 and 1988. Although very large areas were open for exploration, the promising areas in the Campos Basin remained reserved for Petrobrás. The state oil company's dominant position was strengthened further when, in 1988, the programme for foreign participation was closed down and Petrobrás' monopoly for exploration and extraction became part of Brazil's new democratic constitution.

However, with large debt problems, an astronomical inflation and an economic crisis which only intensified in the 1990s, there was a shift in the political winds in Brazil. After managing to control inflation, President Fernando Henrique Cardoso also acquired a political mandate to make a number of changes in oil policy. In 1995, the clause in the constitution which confirmed Petrobrás' monopoly was abolished (Constitutional Amendment no. 9, 1995). In 1997, new legislation which formed the basis for a licensing system with annual licensing rounds open to foreign participation was introduced (Law 9478).

Nevertheless, in 2006, once again an oil discovery made by Petrobrás transformed future expectation for Brazil's oil industry. The development of new drilling technology and seismic equipment that could see significantly clearer and farther into the ground led Petrobrás' geologists to consider the possibility of finding oil under a 2,000-metre thick salt layer which was located in very deep waters in large parts of the Santos Basin. The same deep, oil rich 'pre-salt' layer stretched well into parts of the Campos Basin. When Petrobrás discovered the large Tupi field in an exploration block where British Gas and the Portuguese company Petrogal also had small shares, yet another new chapter in Brazil's oil history commenced. Tupi, which was later renamed Lula, was discovered at a depth of 2,000 metres. The oil lay at 4,850 metres under the ocean floor, meaning that the total depth was some 7,000 metres below the surface.

With fields such as Lula and Libra (2009), Brazil saw the possibility of become self-sufficient in oil for the first time. Hence, once again, oil policy was changed. The pre-salt fields were to be operated by a Production Sharing Agreement-regime (ANP, 2010). Foreign companies could hold limited ownership shares. The formal operative responsibility was left to Petrobrás alone. In the early 2000s, during the development of the discoveries in the deep-water Campos Basin, President Lula and his oil and energy minister, Dilma Rousseff, had gradually introduced stricter requirements for participation by local Brazilian suppliers. After the pre-salt discoveries, from 2011 with Dilma Rousseff as Brazil's president, the so-called local content requirements were tightened further. Petrobrás launched massive investment plans that could lay the foundation for a comprehensive, long-term growth of a local Brazilian supplier industry.

However, the fall in oil prices combined with a corruption scandal in which both Petrobrás and many of the new Brazilian players in the supplier industry were involved, led to the collapse of the ambitious strategy in the autumn of 2014. In May 2016, President Dilma Rousseff was forced to resign. Rousseff was regarded as the architect of the ambitious oil policies. Michel Temer, who took over as president, introduced several measures in the autumn of 2016 where the conditions both for foreign oil companies and foreign oil suppliers emerge as more favourable.

Local content and the role of the suppliers

As part of the protectionist policy that characterised most Latin American countries in the post-war period, from the commencement of the offshore petroleum industry, the Brazilian authorities were concerned with ensuring local Brazilian suppliers a position in the new business. Nevertheless, where Petrobrás faced technological challenges, of which there were many, the company never hesitated to hire the most skilled professionals in the world to be found at the time. From the start in the 1960s and until recently, there was a symbiosis between Petrobrás on the one hand, and the most developed part of the international offshore supply industry on the other. With a few exceptions, the first contracts went to American companies which were either linked to the development of the US oil industry onshore, or the new offshore activities in the Gulf of Mexico.

However, with the strong growth in the technologically challenging North Sea, competence was increasingly brought to Brazil from here. In the 1960s, 70s and 80s, ships, drilling rigs and crews could shuttle not only across the Atlantic between the Gulf of Mexico and the North Sea, but also across the equator between Brazil and the North Sea. Initially, this kind of knowledge transfer took place inside the American supply companies that dominated key areas in both markets. In 1965, the seismic company Western Geophysical Co. was working for BP in the North Sea when, in the following year, crews and equipment were moved to Brazil (Johan *et al.*, 2011). Diving companies such as the American Oceaneering and the French Comex, built on the experience from diving on the Norwegian shelf when they were given contracts in Brazil in the 1970s and 80s. The US drilling companies Sedeco and Nobel drilling, both with many contracts in Brazil, also brought experience from the North Sea (Offshore Frontiers, 2000).

Even after Brazil introduced a strict 'local content' requirement from the early 2000s, foreign offshore expertise remained central for the further development of Brazilian offshore activities. In line with an amendment to the Constitution in 1995, Brazil could no longer discriminate against companies with foreign ownership (Constitutional Amendment no. 6, 1995). The 'local content' policy aimed to ensure that foreign companies carried out as much as possible of the value added locally. In the following years, the requirements became increasingly complex, with various differentiated percentages for various types of fields. From around 2010, the practice was such that all foreign suppliers who wanted to achieve major contracts in Brazil had to use certification companies to confirm the local content proportion in future plans. But even if the local content requirements were costly in some cases, they were not an obstacle to increased imports of provider services or to foreign-owned companies expanding their activities in Brazil. After the first attempts to build up a local Brazilian supply industry nearly collapsed during the debt crisis in the 1980s and early 90s, many locally-owned businesses experienced growth from the early 2000s. But the overall increase in the level of activity was broad enough such that there was room for both increased imports and increased sales by locally-owned as well as foreign-owned companies.

The first Norwegian participation

The first significant Norwegian participation in the Brazilian oil market came as early as 1979 when the Norwegian firm Aker acquired a 75 per cent stake in a yard that constructed legs for jack-up platforms in Aracajú (Akergruppen, 1979). The coastal town Aracajú was located not far from the area where the first discovery offshore was made in Brazil. Aker's commitment in Brazil was a consequence of the optimism linked to the Brazilian oil industry as a result of Petrobrás's many discoveries in shallow waters and the opening of large areas to foreign oil companies by President Geisel.

Aker ran the yard with around 600 employees as a joint venture with the Brazilian group A. Araujo S.A. Foreign participation through joint ventures was a typical means of ensuring increased local activity in the first phase of Brazil's offshore oil operations. Joint ventures were a common instrument in the development of local expertise worldwide in the 1970s. Foreign companies would enter into a teaching role in relation to the local company.

At the time of investment in Brazil, Aker already had come a long way in converting its business from being a pure shipbuilder to become a supplier of semi-submersible floating rigs and large production platforms in Norway. However, the joint venture agreement in Brazil was remarkable in the sense that at the same time as its Norwegian shelf operations, the company was part of a joint venture with the major US construction company Brown and Root, with the American company in the teacher's role. This apparent paradox can be explained by the fact that jack-up structures of the type that would be produced in Brazil were far more primitive than large, complex production facilities of the type being built in Norway. Nevertheless, the initiative was a clear indication of a strong impulse for finding new markets for the Norwegian firm, even in a period when the Norwegian market was growing behind protectionist barriers.

The acquisition in Araujo was, however, not a success for Aker. First, the yard was located far north of the Campos Basin where almost all of the activity was concentrated in the years that followed. The foreign oil companies' eagerness to engage in Brazil was gradually curbed when no substantial discoveries were made. When Petrobrás moved its focus to deep waters, there was also no longer use for jack-up structures which were best suited for shallow water operations. As in Norway and the UK, the crisis in the traditional shipbuilding industry created overcapacity for the kind of simple metal structures the yard could construct. Aker pulled out of Brazil in 1983 (Akergruppen, 1983).

Subsea operations

The development of several deep-water fields in Brazil in the second half of the 1990s presented a possibility for Norwegian companies' newly acquired deep-water expertise. The Baracuda field, which was situated at around 300 metres deep, constituted a technological frontier. Here it was possible to use much of the same subsea equipment that was also used in shallow water, but the equipment had to be operated by sophisticated, costly and risky divers. The largest deep-water diving contract went to the French-owned company Comex (Swann, 2007). Comex had a strong Norwegian link in the sense that the company used diving tables, equipment and labour leaders, previously developed to serve missions in connection with the laying of the Statpipe pipeline down to a depth of more than 300 metres in the Norwegian Trench.

At depths of more than 300 metres, however, where the Marlim field and other large fields was found, all work had either to be performed with the aid of automated equipment controlled from platforms on the surface or by remote-operation vessels (ROVs). To accomplish the necessary services, former diving firms had to be upgraded to become a service industry with a fleet of specialised vessels and advanced ROVs that could install, maintain and repair installations under water. With Norway's strict safety regulations on deep diving and Statoil's and other companies' important long EPC contracts for subsea technology in the early 1990s, Norwegian companies were well placed to assert themselves in all of these areas (Chapter 9).

On many of the Brazilian oil fields in shallow water, equipment such as christmas trees had been provided by the partly Brazilian-owned company CBV Indústria Mecanica. Like Kongsberg Offshore in Norway, CBV had produced equipment under licence for many years from the US company Food Machinery and Chemicals (FMC) (PRNewswire, 1998). However, CBV did not develop its own capacity at the same pace as its Norwegian counterpart. In 1997, FMC strengthened its Brazilian presence by buying up CBV. Since the Kongsberg-part of FMC was currently responsible for the company's activities in the subsea sector, the Norwegian department was central in the purchase.

FMC did not have sufficient capacity to supply all the underwater equipment needed for deep-water operations in Campos. Moreover, Petrobrás was not interested in finding itself in a situation where a single contractor – partly local-based but internationally owned and controlled – had a monopoly on the supply of underwater equipment. This created an opening for Kværner which delivered its first subsea christmas tree in 1997. In 1999, Kværner delivered its first advanced manifold in Brazil. Into the 2000s, Kværner's subsea division continued to have a central position in the Brazilian market. In 2008, now under the name Aker Solutions, the company was awarded a contract delivering 45 christmas trees and related equipment (Offshore.no, 2008). The subsea installations were to be installed at a depth of 2,500 m in the Tupi field.

Kvaerner's first deliveries in Brazil came from a factory in the US that the company had acquired in connection with an attempt to establish themselves in the US subsea market (Chapter 9). When Aker Solutions was awarded the large Tupi-contract in 2008, the company was in the process of setting up a factory at Curitiba in Brazil. Initially, Kvaerner used a converted pulp factory for this purpose. However, when Aker Solutions decided to invest in a brand new top-of-the-art factory in the same area, this was not only due to local content requirements, but was also based on the expectation that deep-water developments in Brazil would be an important market in the future. However, setting up a new advanced production line with necessary upgrading of local sub-suppliers was challenging. In 2012, Aker suffered a loss of some 600 million NOK due to the delays (Offshore.no, 2012a).

Aker's problems strengthened the position of its competitors. In March 2012, FMC got its largest single contract ever when the company was awarded a four-year agreement for the supply of 78 subsea trees the companies facility in Rio de Janeiro (offshore.no, 2012b). The contract corresponded to a turnover of wholly 1.5 billion dollars. When the crises hit and Petrobrás started to scale down its investments plan in 2015, all major suppliers of subsea installations were affected. However, Petrobrás – which had a strong incentive to keep costs down, did not want to place all contracts with one firm. Aker, which had overcome problems with late deliveries, was to a certain degree in a protected position. Nevertheless, with spare capacity, Aker started to bend part of future production towards the deep-sea markets in Africa, by exploring the possibility to become a Brazilian exporter.

Subsea service

Comex retained a central position Brazil also when ROVs replaced divers in deep water. However, when the company was taken over by the Norwegian shipowner Stolt-Nielsen in 1992, its Norwegian connection was strengthened. In the 2000s, the main competitor to Comex (now named Acergy) in Brazil was Subsea 7, controlled by the Norwegian shipowner and entrepreneur Kristian Siem. Alongside advanced diving support vessels and numerous types of advanced ROVs, the large subsea service companies required several advanced pipe-laying vessels. These were vessels with powerful advanced cranes for various types of construction work, in addition to a series of other types of support vessels. Although both Acergy and Subsea 7's 'Norwegianness' was considerably watered down, up until 2010 they both figured on lists of 'Norwegian' companies with significant activities in Brazil. This did not change when the two companies merged the same year, retaining the name Subsea 7.

In 2011, Subsea 7's Brazilian turnover totalled 687 million dollars, with a 2,584 million dollar backlog (Subsea 7, 2011). This accounted for 30 per cent of the company's unrealised contracts. The number of employees in Brazil at this point was around 2000. Several of the future contracts were related to the laying and inspection of pipelines. The largest single contract was an engineering, procurement, construction and installation (EPCI) subsea contract for pre-salt fields Guará and Lula. Three of the vessels to be used,

Seven Oceans, Seven Seas and Scandi Seven, were Norwegian (Stavanger Aftenblad, 2011).

Besides an office in Rio de Janeiro, Subsea 7 had major bases in Macaé and Rio das Ostras, both with storage capacity for large equipment and parts for repairs. To handle the flexible pipes that were essential to connecting subsea structures with platforms and FPSOs, the company had spool bases where pipes were connected before they were loaded on to pipe-lying vessels. Such bases require large sites. Subsea 7's plant north of Rio de Janeiro had a length of 2,225 metres and a land area of 88,000 m^2. After the merger of Subsea 7 and Acergy, only Saipem, a subsidiary of the Italian oil company Eni, and the French company Technip which could rival Subsea 7 on large EPIC contracts offshore.

As with many other companies, Subsea 7 was hit by a combination of the reduced activity in Brazil and Brazil's local content policies. In February 2017, Subsea 7 announced that the contract for the company's Pipelay Support Vessel (PLSV) had been terminated a year ahead of schedule. The company stated that this was due to Brazil's maritime law that gave priority to vessels which were registered in Brazil (Subsea 7, 2017).

Shipyards

Shipbuilding was a central part of the Brazilian local content policy into the 2000s. A number of old shipyards were renewed and extended and new yards were built from scratch. Many shipyards were built with assistance from South Korean and other Asian companies (Intsok 2010, p. 28). Between 2000 and 2011. the number of shipyard workers grew from 1900 to 59,000. Almost all employees were working with offshore-related projects.

With one notable exception, there was no comprehensive Norwegian investment in traditional shipbuilding. Norwegian offshore suppliers at this time were in the process of placing the construction of hulls, large platform structures and production ships out to Asian, especially South Korean, shipyards. A large proportion of the production vessels and drilling rigs that were used in deep-water Campos field were then imported directly to Brazil from Asian yards as well.

The exception was Aker Yard's acquisition of the Niteroi-based builder of the supply-ship Promar in 2001 (Vard, 2014). This acquisition could seem to be a repetition of Aker's former 1979 acquisition in Brazil. However, this was an 'Aker' company growing out the maritime cluster on the north-west coast of Norway, and not the old Aker shipbuilding company. The initiative that led to the building of Promar was taken in 1995 by a small local ownership group that bought up parts of an old, unused yard area previously run by Petrobrás in Niteroi, on the other side of the bay of Rio de Janeiro. The goal was initially to perform simple repairs for a growing fleet of supply-ships. In 1998, the company decided to commence with the construction of simple supply-ships. For Aker, the acquisition of Promar was affected by the company desire to strengthen its general presence in Brazil. This could provide a platform for success in other, more advanced parts of the offshore market. Aker Yard's Norwegian contribution came particularly in engineering. A significant part of the engineering was either carried out in Norway, or by Norwegian engineers who worked from the company's offices at the shipyard in Brazil.

When, in 2008, Aker Yards was sold to the Korean company STX (Chapter 9), the maritime expertise still remained in Norway, soon to be named Vard, and which constituted the real foreign expertise kernel at the Brazilian yard. Vard remained the central unit also after an Italian company acquired ownership in 2013. The connection with the Norwegian maritime cluster also influenced the choice of key sub-suppliers. The most advanced equipment, particularly systems for dynamic positioning and propellers, were supplied by Rolls Royce's Norwegian division which was both geographically and technologically located at the core of the Norwegian maritime cluster. Regardless of ownership, as long as engineering work was done by the Norwegians, who clearly had a preference for specialised equipment that they knew and trusted, they contributed to giving other smaller Norwegian suppliers an advantage in this market.

While Vard was the only Norwegian company that took part in the construction of ships, the number of Norwegian companies that took part in the growing supply and service marked in relation to the increasing number of Brazilian offshore-related yards was far greater. The equipment and services needed for a yard that was constructing supply-ships was, of course, different from a yard constructing drill-ships, semi-submersible drilling rigs, or undertaking the final installation on floating production storage and offloading vessels (FPSOs). For most types of vessel that could be moved, dynamic positioning system was a necessity. Most types of large installation were also dependent on the different types of cargo-handling equipment. This was equipment that had to be installed before the installations were moved out to the oil fields. A company like Aker Solutions carried out activities related to several of these sub-areas simultaneously. Other companies were geared towards more specific areas.

A significant proportion of the floating rigs, exploration ships and production facilities that were imported into Brazil from Asian shipyards often had equipment from Norwegian suppliers on board. This could ensure companies which had supplied this equipment in the first place, contracts related to maintenance, repairs and modifications when the installations were operating in Brazil. So-called 'drilling packages' comprised the single area where Norwegian suppliers in the 2000s had the most dominant position in the offshore market. Aker and National Oilwell Varco (NOV), which was American-owned but where the Norwegian branch was responsible for deliveries, were the dominant firms worldwide. When, in the years just before the crisis, Brazilian shipyards developed the capacity to produce advanced offshore installations on their own, both NOV and Aker Solutions were awarded substantial contracts. In 2012, 35 rigs operating offshore Brazil had drilling equipment from Aker Solutions (Aker Solutions, 2012). Most of these rigs were built in Asian shipyards. The same year, however, Aker acquired a contract to supply drilling packages for six drill-ships to be built by Estaleiro Jurong Aracruz shipyard in Brazil (Aker Solutions, 2012).

The supply vessel market

Although Brazil started to manufacture some supply vessels from the end of the 1990s, it was not possible to meet the demand created by the tremendous growth in activity offshore in the 2000s. The new growth demanded also a number of more specialised vessels that had not previously existed in Brazil. This created an opening for Norwegian supply-ship owners. From the early 2000s until the Brazilian oil boom reached its peak around 2013, Norwegian supply-ship companies such as DOF ASA, Farstad Shipping, Solstad Offshore, Siem Offshore and Eidesvik Shipping expanded extensively in the Brazilian market. Norwegian companies had become the second largest foreign nation in the market. At some point during the autumn of 2012, Norwegian ship-owners operated 50 Norwegian vessels in Brazil. Farstad Shipping, the largest Norwegian company in this market, had a turnover of 243 million NOK in Brazil in 2011. This was equivalent to 27.9 per cent of the company's total turnover. With offices in both Rio and Macaé, the company had 13 vessels which were active in Brazil.

Brazil became an important driver in the internationalisation of that part of the Norwegian offshore supply market which has traditionally had a relatively national approach. The Norwegian supply-ship owners retained strong Norwegian roots in the sense that a considerable number of the vessels they operated in Brazil were built in Norway. But despite its entrenched Norwegian background, the supply-ship owners had learned from other parts of the shipping industry to utilise the advantage of being able to sail under international flags and create company units that made it possible to avoid taxes and certain regulations.

Moreover, the Norwegian supply-ship owners adapted well to the Brazilian requirement that two-thirds of the crew on board vessels operating in Brazilian fields should be Brazilian. One might expect that there were economic benefits in using cheap local labour. Here, however, the traditional notion of multinational oil companies that exploit cheap local labour was turned upside down. Because the Brazilian navy had a monopoly on training maritime personnel, there was a considerable lack of relevant competent local workforce. This led to higher wages for Brazilians than for Norwegians. When the company, Eidesvik, considered pulling out of Brazil, it referred to the fact that wages for local employees were around 20 per cent higher than for Norwegians (Stavanger Aftenblad, 2012). The differences were so great that Norwegian ship-owners preferred experienced Norwegian crew in key positions, although these commuted back and forth between Norway and Brazil at the company's expense.

The rig market

When Norwegian offshore rig companies became engaged at a relatively late stage in Brazil's oil boom in the 2000s, this was due to the fact that the Norwegian fleet had been scaled down considerably during the period of low oil prices in the late 1990s. But with the creation of John Fredriksen's Seadrill from 2005, capacity was greatly expanded. The family-owned company, Odfjell, which had been very careful not to engage internationally in the 1990s, similarly increased its capacity by ordering new large rigs that could be used in deep water in Brazil.

In 2008, Seadrill had its breakthrough in Brazil with a long-term contract with Petrobrás for deep-water drilling. This contract was referred to as 'the all-time contract' by the Norwegian newspaper Dagens Næringsliv (2008). It was in many ways an apt description. Never before, neither on the Norwegian shelf nor elsewhere, had a Norwegian rig company signed a contract that had a similar value. Seadrill would operate three deep-water rigs with an average daily rate of 624,000 dollars over a six-year period, from 2009 to 2015. Income potential throughout the period was for a total of around three billion dollars. Seadrill would use the three semi-submersible rigs – West Eminence, West Orion and West Taurus. Two of the rigs were to be built at the Jurong shipyard in Indonesia, while the third was to be built at the Samsung shipyard in South Korea. The rigs, all of which were built in accordance with the Norwegian technical standards and therefore had much Norwegian equipment, were sent to the other side of the globe to Brazil, as soon as they were completed in the Asian yards.

Odfjell Drilling secured its first contract in Brazil as late as 2011. The company would operate the drill-ship, Deep Sea Metro 2, where the Norwegian company had a 40 per cent stake. In August 2012, Odfjell expanded its Brazilian activities when a potentially important contract with the Brazilian partners Galvão Óleo and Gás and SETE Brazil was signed. Together these companies ordered three new drill-ships to be built in Brazil (Sysla, 2012). Odfjell would have a 20 per cent stake in the three ships. The company was to operate the vessels through a joint venture with Galvão in the first operating period of five years. This contract was related to SETE Brazil where Seadrill and the Norwegian-Cypriot company, Ocean Rig, were also involved (offshore.no, 2012 c).

In 2015, both Seadrill and Odfjell had to scale down their activities in Brazil to a minimum. Low oil prices led to a particularly sharp decline in drilling activity; moreover, Brazil's corruption scandal contributed largely to paralyse SETE Brazil.

PGS and BW Offshore

The two largest Norwegian suppliers in the Brazilian market not mentioned so far are Petroleum Geoservices (PGS) and BW Offshore. With its strong international orientation from its establishment in 1991, PGS acquired its first contract in the Brazilian fields in 1994 (PGS, 2009). With its large modern seismic vessels and access to very large computer capacity to analyse results, there were no local alternatives. PGS's activities in Brazil reached a peak between 2009 and 2011, with a turnover of around 200 million dollars annually (PGS, 2011, 86). This amounted to between 15 and 20 per cent of the company's total annual revenue. Typical for the offshore oil industry cycle was that the seismic segment was hit first when curves turned downward. PGS was therefore already in the process of winding down its activities in Brazil when the corruption scandal broke in 2014. During the 2000s, BW Offshore had established a strong position in an emerging market for the construction, rental and operation of FPSOs.

The sub-suppliers

The explosive interest in the Brazilian oil industry in the pre-salt boom from around 2008 was decisive when a number of smaller Norwegian subcontractors decided to invest in Brazil. Small Norwegian subcontractors' path into the Brazilian offshore market could be different, based on which sub-segments of the industry they were operating in. For many companies, the Norwegian government and government-supported industry organisations were more important for the establishment of smaller companies than for larger companies (Interview, Vik, 2013).

The crucial support institutions were in place in Brazil before the 2008 superboom. Similarly to other countries with significant Norwegian commercial interests the Norwegian Brazilian Chamber of Commerce (NBCC) was established in 1995. The NBCC gained critical mass as an increasing number of Norwegian actors became established in Brazil. A number of venues where the smaller Norwegian players could connect in order to become familiar with relevant Brazilian actors were established. From 2003, Innovation Norway, a government sponsored institution for supporting Norwegian businesses, had an office associated with the Norwegian Consulate in Rio de Janeiro. Innovation Norway collaborated with Norwegian Oil and Gas Partners (INTSOK), which also had its resident representative in Brazil. Nevertheless, the main gateway for smaller suppliers was the relationships they already had with larger Norwegian EPC suppliers and foreign companies operating in Norway. Moreover, Norwegian subcontractors had an advantage in knowing the technical standards their traditional customers operated by. The establishment of a small Brazilian subsidiary of the Norwegian company, Kleven, is a typical example in the sense that several of the conditions referred to above were affected simultaneously.

Kleven was a small shipyard based in the maritime cluster in western Norway already mentioned. The establishment in Brazil was based on a subsidiary that delivered electrical services to Kleven's yard Norway. The aim of the establishment was to provide electrical services to the large supply-ship fleets operating in Brazil. Kleven established a small factory in the same industrial area in Niteroi where Vard also had its yard. In the bay between central Rio and Niteroi 50 supply vessels or more lay anchored, at any time. A large proportion of these were Norwegian.

Time during anchorage was often used for maintenance and repairs. However, the Norwegian ship-owners who operated the ships were dissatisfied with the local electricians who were not familiar with the security restrictions on a marine vessel that operated close to oil installations. The contact between

Kleven and the local Brazilian partner was taken up at one of the many events organised by the Norwegian support network in Brazil (Interview, Vik, 2013). When the company was well-established in 2013, it had a management consisting of a Norwegian, a Dane and a Brazilian. These led a group of about 30 local electricians. The local electricians were trained in the standards that were required on Norwegian ships.

Another somewhat similar example is the container company, Modex Energy (O'Hanlon, 2015). On the Norwegian and British continental shelves, the Norwegian certification company, Det Norwegian Veritas (DNV), defined specific technical standards for the use of containers offshore. This resulted in containers which were in use becoming significantly more advanced than containers used on trailers, trains and carriers worldwide. Typically, in Brazil in the early phases, traditional containers, provisionally split into two separate parts, were in use offshore. With the upgrading that took place in the in the 2000s, Petrobrás adopted maritime standards developed by DNV. Hence, a market for subcontractors which were familiar with the DNV standard was created.

Modex was not the first foreign container company that established itself in Brazil. The leading company was the British Swire group, competing in the same market in the North Sea. Since Petrobrás wanted competition in this small limited market, Modex was virtually invited to Brazil. When the container industry was also hit by the crisis in 2014, this turned out to be advantageous to the smaller of the two main competitors. Although the overall market declined, Petrobrás had an interest in maintaining a certain competition. Thus, it was the larger of the two actors which had to reduce their deliveries most.

Conclusion

Despite difficulties in the aftermath of the fall in oil prices commencing in 2014, Norwegian participation in the Brazilian offshore market has been a success story. Since Brazil and Norway opened up for oil operations at the same time, and for long periods has had a parallel development, Norwegian companies' success in Brazil indicates that from the early 1990s the Norwegian supply industry had reached a level where a number of companies were leaders in various sub-segments of the industry. The essential background for the success was the expertise developed in the 1970s and 80s, the first two formative decades, leading up to an internationalisation of the industry (see Chapter 7).

The first competence transfer from the Norwegian to the Brazilian oil sector occurred in the early years through the same, mostly US, companies which dominated as suppliers of the most advanced technology. When, as early as 1979, Aker bought up a medium-sized shipyard in Brazil, this confirms that from an early stage there was a significant international orientation among key Norwegian offshore supply companies.

With Petrobrás as dominant operator, Brazil was different from many other offshore markets in the sense that Norwegian suppliers were hardly able to take advantage of the relationships with major foreign oil companies that were established through contracts on the Norwegian shelf. When the first breakthrough for Norwegian suppliers in Brazil came in the subsea segment at the end of the 1990s, it was as that part of the international offshore industry where Norwegian companies were most advanced technologically. The subsea expertise of companies such as FMC and Kvaerner (later Aker Solutions) was developed in relation to the type of EPC contract as especially Statoil, but also other oil companies with operatorship on the Norwegian continental shelf, started to use in the early 1990s. When Petrobrás sought the same expertise for the development of deep-water fields in Brazil from the late 1990s, the Norwegian suppliers had both the technological expertise and organisational capacity that was required.

Whereas Aker's first, unsuccessful, engagement in Brazil was related to a joint venture, the period commencing in the 1990s was characterised by a far wider range of organisational solutions for Norwegian firms. Many Norwegian owners bought up local Brazilian companies, and business expanded on the basis of these. In some cases, Norwegian companies established production facilities from scratch. In the early years, many old factory buildings and factory areas were redeveloped for offshore-oriented purposes. In other cases, such as Aker Solutions in Curitiba, completely new production facilities were established.

In the 2000s, when Norwegian companies also became established in segments such as supply vessels, drilling, drilling packages and other specialised operations, technological skills were crucial for success. For the biggest companies, however, success in Brazil reflected the fact that they had established a strong position in what could be described as global production chains. In being technologically competent, well-integrated into an increasingly globalised industry and with good access to capital, Norwegian industry had the necessary advantages to expand quickly as the Brazilian market required. Finally, a key both to large as well as smaller Norwegian suppliers was their knowledge in dealing with international standards.

References

de Oliveira A. (2011). Brazil's Petrobrás: Strategy and Performance. Victor, D. G., Hults, D. R. and Thurber, M. C., *Oil and Governance: State Owned Enterprises and the World Energy*. Cambridge: Cambridge University Press.

Johan, P., Abreu, C. E., Grochau, M. and Thedy, E. A. (2011). Advanced Seismic Imaging Impacting Brazilian Offshore Fields Development. Presentation at Offshore Technology Conference, Houston, Texas OTC 21934.

O'Hanlon, J. (2015). Containers and cabins with conviction. *Energy Digital*. Retrieved 23.02.2018.

Priest, T. (2016). Petrobrás in the History of Offshore Oil. Schneider, B. R. (ed): *New order and progress. Development and Democracy in Brazil*. Oxford: Oxford University Press.

Randall, L. (1993). *The Political Economy of Brazilian Oil*. London: Praeger.

Ryggvik, H. (2018). The offshore industry in Brazil: A Norwegian Historical Perspective. Unpublished manuscript.

Swann, C. (2007). *The history of oilfield diving*. St Barbara: Oceanaut Press.

Additional sources

Aker Solutions (22.06.2012). Aker Solutions bygger ny fabrikk for boreutstyr i Brasil [Aker solution builds new factory for drilling equipment in Brazil]. (Press Releases).
Akergruppen. (1979). A/S Aker Mek. Verskted, Årsrapport. (Annual Report).
Akergruppen. (1983). A/S Aker Mek. Verksted, Årsrapport. (Annual Report).
ANP legislation for Oil and Gas Exploration of oil and Natural Gas. (English translation of Law 12276 of 20 June, 2010; Law 9478 of 6. August, 1997; Law 12734 of 2 August, 2010; Law 12351 of 22 December, 2010.)
Constitutional Amendment no. 6. 15. August, 1995.
Constitutional Amendment no. 9. 9. November 9, 1995.
E24. (25.10. 2016). Britene kjøper mest petroleumsteknologi [The Brittains buy the most petroleum technology]. (Newspaper article).
Interview with Jan Arild Vik, Kleven ORN, Niteroy. (23.10.2013).
INTSOK, Shipyards in Brazil. (06.2010).
Offshore Frontiers (08.2000). Transocean Sedeco Forex First in Brazil.
Offshore.no. (18.12.2008). Aker Solutions første dypdykk på Brasilianske Tupi. [Aker Solutions first deep dive on Brazilian Tupi].
Offshore.no. (21.03.2012b). Aker tapte 600 millioner i Brasil [Aker lost 600 million in Brazil].
Offshore.no. (10.02.2012c). Booker 26 rigger i 15 år [26 new rigs booked in 15 year].
Offshore.no. (19.03. 2012a). Gigantkontrakt til FMC [Giant contract for FMC].
PGS. (2011). Annual Report.
PGS. (16.04.2009). Another Valiant Performance in Brazil. www.pgs.com.
PRNewswire (14.08.1998). FMC acquires CBV Industria Mecanica.
Siem Offshore Annual Reports (2012-). Retrieved 28.02.2018 www.siemoffhsorereports.no.
Stavanger Aftenblad. (7.04.2011). Kjempekontrakt til Subsea 7 i Brasil (Huge contract to Subsea 7 in Brazil).
Subsea 7 S.A. Annual Report. (2015).
Subsea 7, Annual Report. (2011).
Subsea 7, News. (16.01.2017.) Subsea 7 announces early termination of contract offshore in Brazil.
Sysla 6. (08.2012), Petrobrás hand out contracts for 12 rigs.
Vard, a Fincantieri Company (11.2014). Powerpoint presentation.

9 Supply companies and the political economy of platform concepts in the U.S. Gulf of Mexico

Helge Ryggvik

Introduction

The oil industry has been markedly international ever since the start of the twentieth century, in that the largest oil companies established themselves wherever large quantities of oil were to be found. But even though several of the major colonial powers established their own oil companies for strategic reasons, much of the technological expertise that was decisive for discovering and producing oil was found in the US. US dominance was further strengthened from the 1950s when the oil industry took its first serious steps offshore in the Gulf of Mexico. American oil technology and oil expertise were thus, as we have seen, entirely central in the search for oil in the North Sea and in Brazilian coastal waters from the 1960s.

Between the 1970s and the turn of the millennium, several radical changes affected that part of the industry which was oriented towards offshore oil activities. Many American supply companies held a strong position, even if the volume of production in the Gulf of Mexico did stagnate until the early 1990s. On the basis of the extensive offshore market in the North Sea, competitive supply companies developed, particularly in Norway (see Chapters 2 and 7), Great Britain and France, Italy and the Netherlands. Moreover, from the early 1990s both the European and American parts of the supply industry experienced significant globalisation (Odemis, 2015; Ryggvik, 2013). Product chains were established where the engineering, construction and completion of equipment could take place across three continents. At the same time, genuinely globalised organisations were established, creating close connections between technological communities in different states and regions, and where in one and the same company exports might go in all directions.

Starting from this position, one might have expected that technological developments from the 2000s on would continue along broadly similar lines within a continually growing international market. However, if one compares the development in what would remain the three largest offshore regions (North Sea, Brazil and the Gulf of Mexico), there are nevertheless remarkable differences. Some of these differences can be ascribed to different geographical and geological conditions, but others can hardly be explained without including historical and social factors. These differences are expressed both in different technological choices and in different ways of organising markets.

This chapter explores the political economy which has shaped the development of major fields in the US Gulf of Mexico. Where relevant, we note corresponding technological choices in Norway and Brazil.

Technological style on the deep sea

No other parameter has greater implications for supplies to an oilfield than the design used to develop the field. At one level, the choice of a field development design is concerned with finding an appropriate way of combining existing technologies; but developing an oilfield, where geological and geographical conditions are almost always different, involves so many different complex considerations that the solution arrived at constitutes an innovation in itself. Moreover, the main design from which one starts requires adapting other technological elements, the sort of adaptations which can drive further innovations. Key technologies such as christmas trees, manifolds, and underwater pipelines have to be adapted every time, both in relation to the quality of the subterranean oilfield, the physical conditions affecting the water depth in question, as well as what types of installation are to receive the oil or gas on the surface.

As early as the mid-1970s, clear regional distinctions in field design had developed in the sense that large, fixed concrete installations were used extensively in the Norwegian sector, while the rest of the world mostly used the steel chassis. Olsen and Engen have shown how this sort of Norwegian 'technological style' was not merely an engineering solution to the geological and geographical challenges of the Norwegian continental shelf, but also expressed a 'path dependency' affected by the particular characteristics of local industries, which in turn contributed to sustain a particular supply model (Engen and Olsen, 1997). This path dependency was also sustained by the protectionism which characterised the oil market in the North Sea from the mid-1970s (Chapter 9).

If it was above all the choice of chassis materials that constituted a clear dividing line in the choice of offshore development solutions up to the late 1980s, the deep-water revolution meant that as from the 1990s there was a range of very different options available for developing an offshore oilfield. In the case of Norway, social conditions still affected technological development, not least a strong regulatory regime which emphasised robust solutions that made extensive use of technical safety barriers. At the same time, it might seem that the removal of protectionist barriers and the subsequent internationalisation of the supply industry contributed to detaching the dominant operators from their previous path dependency. Commencing in the 1990s and into the 2000s, a wide spectrum of development solutions was used (Chapter 2). The same was true in the British sector. However, in the two other major offshore markets, the US and Brazil, there were clear signs of a path-dependent dynamic recalling that which previously affected Norway.

Unlike the British, and even more so the Norwegian sector of the North Sea where the conditions were very different from those in the US, the depths and geological conditions in the Gulf of Mexico and Brazilian coastal waters were relatively similar. It is thus all the more noteworthy that Brazil and the US Gulf of Mexico ended up with very different 'technological styles'. While Norway in particular used several different solutions from the early 1990s, the US and Brazil acted as polar opposites around major technological choices. While developments in Brazil, as we have seen (Chapter 10), were dominated by extensive use of production ships or Floating Production, Storage and Offloading vessels (FPSO), deep-water developments in the US sector of the Gulf of Mexico were characterised by a more complex picture of different types of floating production installations without storage capacity.

FPSOs had already made a definitive breakthrough as a development concept in the international oil industry in the 1990s. By 2015, a total of 39 production ships were operating on the Brazilian continental shelf (Wood Group Mustang, 2015); 38 FPSOs were producing in West Africa, while the numbers for the North Sea, Southeast Asia and China were 23, 24 and 17 respectively. However, in the USA only one FPSO was in operation at this point. This had only entered production in 2012; tellingly, it was hired by the Brazilian oil company Petrobrás from the Norwegian company BW Offshore.

The US Gulf of Mexico, for its part, was distinctive in that in 2015 a total of 18 giant installations were using the so-called Spar technology. With the exception of a loading buoy in the North Sea (the Brent Spar), which became world-famous because of the large-scale environmental protests when it was due for disposal in 1998 (Parmentiier, 1999), this technology was not used as a development concept in deep-water fields anywhere else in the world until Statoil and Norway chose a corresponding concept for the development of the Aasta Hansteen field, which was planned to be ready for production from 2017 on. Furthermore, the US used a far greater number of so-called Tension Leg Platforms (TLP) than anywhere else.

The similarity between Brazil and the US Gulf of Mexico is that both countries have a large, relatively shallow and broad continental shelf which then falls off to great depths. For the US Gulf of Mexico, a shelf up to 50 metres deep stretches 100–200 kilometres. It is widest off the border between Texas and western Louisiana, and narrowest off the mouth of the Mississippi. Outside this shelf, the ocean bed quickly falls away to 1,500 metres and more. Just as off Brazil, the Gulf of Mexico also has a salt layer, and from a comparably old geological era. However, this layer was not as thick and continuous as in Brazil's Santos basin. At various times, oil companies had drilled very deep here as early as the 1990s, but without seismic technology they could not 'see' down to the actual layers, and drilling was thus guesswork.

However, with new 3D seismic technology from the early 2000s, together with advanced drilling technology, it became possible to aim much more precisely towards possible deposits in deeper layers. Both in the US and Brazil, therefore, there were cases where one and the same installation collected oil from petroleum-bearing layers on different 'levels', at depths differing by as much as 2,000 metres. The parallel between developments in the US and Brazil was demonstrated by how the two countries alternated in holding the record for the deepest drilling and the deepest producing field. For part of the 2000s, most of the records were broken in the US. In 2008, Shell found the Tiber field at 2,800 metres water depth, but which involved drilling down to geological structures lying 12,100 metres under the water surface: the total depth drilled was thus 10,685 metres! (Oil & Gas Eurasia, 2008). In November 2017, Shell set another world record when it started production from the Stones field at a water depth of 2,900 metres in the Gulf of Mexico (Shell, 2016).

Although there were many similarities between the US Gulf of Mexico and Brazil, there was one important geological difference which created historical parameters with technological consequences. In the Gulf of Mexico, there was considerably more oil to be found close to shore than in Brazil. When oil and gas were later found in deep waters, the industry thus could be connected to an extensive infrastructure of existing platforms and, especially, pipelines. With access to these pipelines, connected in their turn to one of the world's largest concentrations of refineries, the need for storage capacity out on the fields was not as great as it was elsewhere.

Four different production concepts in particular would dominate the development of deep-water fields from the 1990s and into the 2000s: TLPs, Spar platforms, floating semi-submersible production units (FPS/FPUs) and FPSOs; all these main categories had their own subcategories. As long as the installations were not already overloaded with production and drilling equipment, all these types could be connected to nearby fields that were operated by underwater installations, a connection designated Subsea tie-back (SSTB). Even if the industry consisted of suppliers which could increasingly deliver goods and services linked to a wide range of offshore technologies, the design of a platform still had a major effect in terms of which companies had the skills and capacity needed to secure the relevant contracts.

Shell and TLPs

In the course of the 1980s, companies such as Conoco, BP, Mobil, Amoco, Oryk and Exxon found oil in deep waters in the Gulf of Mexico. But in the pioneering years, just as in the North Sea with Norway's Troll field, it would be Shell and its contractors who took the lead in the developments that shaped the transition from shallow to deep waters in the Gulf of Mexico. Shell invested most, and was the first company to develop fields deeper than the critical level where divers could operate. When, in 1988, Shell installed a steel jacket at a depth of 412 metres on the Bullwinkle field, there was a general recognition within the industry of having reached the limit of how deep one could operate with fixed installations. The giant steel structure that carried the platform was the world's largest of its kind.

When Shell came to design the production solution for the Auger field which was discovered in 1987 at a depth of 872 metres, the company chose to use a TLP. A TLP could contain the same types of drilling and processing equipment as fixed platforms, but consisted of a floating steel structure anchored to the ocean floor by steel tendons. Unlike FPSOs which moved with the waves, but like installations that stood on the ocean floor, a TLP could use fixed risers in steel. Depending on the drilling equipment on the platform, it was thus possible to carry out different types of intervention in wells during the lifetime of a field. The same functions as a christmas tree, which had valves and channelled the flow in and out of the well, could be placed on the platform. This meant that one could operate without the kinds of advanced underwater trees and flexible pipes required for producing from production ships in Brazil. Fixed production pipes in steel, and the use of dry subsurface christmas trees, were decidedly cheaper. When the choice of concept for Auger was made, there was still some uncertainty around the use of christmas trees that could be used without divers at great depths. Even if TLPs could operate at great depths, then, the most advanced technology remained on the surface, on the platform itself.

The development of the TLP concept was an example of how the development of offshore technology had become internationalised. The first TLP was used by the American oil company Phillips on the Hutton field in the British sector. The Hutton field started production in 1984. This time, however, it was only 147 metres deep. With the Snorre field in the Norwegian sector, which started production in 1992, the same principle was used at 335 metres depth. To triple the depth by comparison with Snorre would require a considerable development of the technology. The platform had to have the capacity to 'carry' the weight of tendons, production pipes, production drilling equipment and other equipment which was necessarily longer, and hence heavier, the greater the depths in question.

Besides the uncertainty around wet christmas trees and the advantage of having constant access to drilling capacity, it was particularly the question of how to transport the oil to land that decided Shell against using a development solution involving storage capacity, which would have corresponded to the solutions using FPSOs in deep water in the Campos basin in Brazil.

The Auger field was about 230 km (137 miles) from the coast of Louisiana. Shell chose a solution where the oil, once processed, was piped on towards land. If it had been necessary to create a new pipeline for this whole long stretch, this would have been very expensive. Here, however, Shell could use the advantage of the extensive pipeline network in the shallow part of the Gulf of Mexico, not far away.

The contractual strategy Shell chose for developing Auger was a complicated one (Enze *et al.*, 1994). When the first contracts were signed in 1990, a number of companies were involved. The project was one of the first large installations in the Gulf of Mexico where the hull was constructed on another continent and then towed to the vicinity of the field for completion there. The hull was built by Bellelli S.p.a. in Italy. From there it was floated and towed across the

Atlantic Ocean on a barge. However, the deck, which constituted a much larger contract, was built back in the US by McDermott Inc. at their Amelia, Louisiana facilities. In addition, numerous other smaller modules were fabricated separately from the deck structure by seven different contractors, precommissioned as a package, and installed on the deck. The tendons were fabricated by the Gulf Marine shipyard in Corpus Christi, at the time owned by the Norwegian firm Aker. McDermott Inc. was responsible for installation of the TLP, which was ready for production in 1994. Thus most of the value creation associated with the project still took place in the US.

On the basis of its experience with the Auger field, Shell followed up throughout the 1990s and into the 2000s with a series of similar TLP developments (Oil & Gas Journal, 1999, Wood Group Mustang, 2015): Mars 1996 (894 metres, 700 million barrels), Ram Powell 1997 (980 metres, 250 million barrels), Ursa 1999 (1,158 metres, 400 million barrels) og Brutus 2001 (910 metres, 200 million barrels). Shell was not alone in using the TLP concept. From the early 2000s, companies such as ATP Oil & Gas, PXP (Plains Exploration & Production), Chevron and Conoco-Phillips also started production based on TLP platforms. The general opinion among oil engineers was that the TLP concept was particularly suited to depths of around 1,000 metres which characterised many developments in the Gulf of Mexico around the turn of the millennium. For Shell's part, however, it was also clear that the company was using the economies of scale involved in repeating a tried-and-tested solution. It was to a large extent the same supply companies who were granted the largest contracts (Offshore Technology, 2017).

But despite the fact that producing oil and gas from TLP platforms contributed to increasing production in the US Gulf of Mexico, several serious incidents in the 2000s raised the question of safety around the TLP concept. In 2004, the Norwegian Snorre A platform came close to sinking when a gas leakage under the platform led to it losing much of its buoyancy (Petroleumstilsynet, 2004). In the dramatic 2005 hurricane season, several platforms suffered serious damage. Shell's Mars platform was badly damaged after Hurricane Katrina in late August (New York Times, 2006), while Chevron's Typhoon platform capsized during Hurricane Rita, which hit the Gulf of Mexico barely a month later (Houston Chronicle, 2005). TLPs would also be constructed for deep-water fields in the Gulf of Mexico subsequently. But once developments reached depths considerably deeper than *c.*1,000 metres and more, in the early 2000s, it was the so-called Spar concept that would become the dominant development concept for the largest deep-water fields for some time.

Spar technology

While Shell had pioneered the use of TLPs, companies such as Chevron, Andarko (KeerMcGee (Oryx)), Murphy, BP and Plains Exploration & Production (PXP) had already used Spar platforms for their offshore deep-water fields from the end of the 1990s. The Spar concept was different from all other offshore installations in that the platform deck was mounted on a single, floating cylindrical hull. The hull consisted of several discrete elements which could contain both smaller storage tanks, floating chambers or ballast to ensure stability. Waves would be able to move up and down the construction as if the platform was standing on the ocean floor. Since a Spar platform's centre of gravity was below the water surface, it was more stable than a TLP.

But the deep-water regions far out in the Gulf of Mexico were characterised in many places by strong, irregular, vertical and horizontal underwater currents that could create 'heave'. This was partly resolved by installing 'fins' in a spiral around the platform hull. The idea was that these streams would push the platform up rather than sideways. Hence, like TLPs, Spar platforms could also use fixed risers in steel. This meant that Spar platforms could also operate with 'dry' christmas trees on the platform. However, a Spar platform's anchor system weighed less than a single TLP tendon. Thus, even if a TLP could operate at great depths, the weight of 'tendons' became an increasingly significant operating challenge the deeper one came. There was thus a physical, and hence an economic, inducement to use the Spar concept once one came to waters significantly deeper than 1,000 metres.

Oryx Energy was the first oil company in the Gulf of Mexico to use a Spar platform as a production concept. The company was the operator on the Neptune oilfield at 588 metres depth where production started in 1996 (Wood Group Mustang, 2012). While this was deep, it was thus considerably shallower than the fields where Shell first introduced TLP platforms. The Neptune platform did not have the capacity to carry out production drilling, but Chevron's Genesis platform, which started production in 1999, did. ExxonMobil took a provisional record for greatest depth when a Spar platform for the Hoover/Diana field was installed at 1,443 metres in 2000.

A Spar platform's diameter and its depth underwater were decisive for its operation, and thus for how much weight in the form of deck and equipment could be placed on the platform. The Holstein platform, installed for PXP at 1,343 metres depth in 2004, took a provisional record for carrying capacity.

The fact that the first Spar platform was installed at a rather shallower depth than Shell's first TLPs shows that the platform design was not just a choice based on physical laws, but also that engineers in the oil companies and the suppliers were proceeding by trial and error. The Spar concept represented an innovative solution which clearly broke with earlier traditions in the oil industry. The idea was developed by the American engineering company, Deep Oil Technology, which held the patent on the Spar concept, and the company's leader, Edward E. Horton, is reckoned as the inventor of the Spar concept (Ocean Star Museum, 2008). When Oryx ordered its first Spar, the ownership of Deep Oil Technology was taken over by the Finnish Rauma Offshore, which in turn was tied to the Mäntyluoto Oy shipyard in Pori, Finland. The Finnish shipyard obtained the contract to build the hull for the Neptune platform.

The fact that the Spar concept was developed by a community slightly separate from the dominant suppliers and then developed further and ordered by

Oryx Energy, one of the youngest and smallest companies to operate in deep waters, might be a confirmation of the proposition that actors independent of established structures can often pave the way for new ideas. Shell's repeated use of TLP, similarly, could illustrate how a company which had initial success with a given concept can exploit economies of scale by keeping to a similar solution in future developments. But the fact that at an early point both Chevron and ExxonMobil chose the Spar solution for developing new fields in deeper waters shows that the barriers to choosing entirely new concepts were not very high for major companies. In 2010, Shell took another new provisional record when it started production from a Spar platform on the Perdido field at a depth of 2,383 metres.

Just as the hulls of TLPs were built in Italy, so the building of Spar platform hulls in Finland showed that there was an opening for foreign actors in the American offshore market. In 1995, before the Spar concept had made its full breakthrough, Deep Oil Technology and Rauma Offshore were bought by the Norwegian Aker company. Aker Rauma carried out engineering work for the hull on the next three Spar installations. For these, too, the hulls were constructed at the Finnish shipyard. In 2000, however, Aker sold its production chain for Spar platforms to its competitor, Coflexip. Shortly afterwards, Coflexip was taken over by the French company, Technip. For Technip, the purchase proved to be a good deal: the company would be a dominant supplier of Spar installations for many years to come.

Parallel to Technip's takeover of the tried-and-tested Finnish-American production chain, the J. Ray McDermott company sought to establish itself as a competitor by securing access to shipyard capacity in the UAE. The company secured contracts from the Murphy oil company to build Spar platforms for the Medusa and Front Runner fields. However, the attempt nearly broke this traditional American engineering and construction company (McDermott, 2002).

First, low oil prices around the year 2000 meant that all offshore suppliers were under severe price pressure when signing contracts. There were a series of problems both in the hull production process and in connecting the hull to the topside. This meant that Technip and the Finnish yard had a near-monopoly position in the years that followed. In 2013, the shipyard in Finland had built 12 of the 17 Spar platforms in international use (Offshore Magazine, 2013).

This dominance was noteworthy. Both on the Norwegian and the British continental shelf at this time, standardisation driven by state authorities and a certain determination on the part of the dominant operator companies combined to ensure that there were always a few suppliers capable of competing for major EPC contracts. Although there was a far larger number of operative oil companies in the US Gulf of Mexico, there was no corresponding indirect regulation of the offshore market. Even though the hull of a Spar platform was more technologically complex than that of a TLP, topsides, different types of drilling and processing equipment on the platform and underwater equipment still constituted the greater part of contracts, whether by cost or in terms of technical complexity. For all the Spar platforms installed in the Gulf of Mexico up to

2016, the topside and significant parts of the equipment on board in the hull were constructed and installed in the USA. Thus, developments here followed the same pattern as for TLPs. A Spar construction was well suited for long-distance transport long as it was in a horizontal position. However, topside and equipment could only be connected once the hull was submerged in the locality where the platform was to operate. The construction of Chevron's Neptune and Genesis platforms is typical here (Oil & Gas Journal, 1996).

Moored semi-submersible systems

More so than other companies, it was BP which would become the pioneer in introducing semi-submersible platforms as production hubs in the Gulf of Mexico: or more precisely, the concept was introduced by the Na Kika platform, which started production from 2002. Na Kika was originally developed by Shell, but BP later become the majority owner and operator. Many in the industry were surprised that Shell did not choose TLPs for the Na Kika hub: 'We like TLPs,' said Dave Lawrence, Vice-President of Exploration and Development for Shell E&P when the project was introduced (Offshore Magazine, 2000). Thus it is a little surprising the company went in a different direction. However, he added that from a cost standpoint it would not make much sense to put a TLP in these water depths. As early as the turn of the millennium, it was generally accepted in the industry that TLPs had reached their economic limit at water depths of about 1,300–1,600 metres.

The Na Kika platform was to produce from a series of associated fields lying around depths of 2,000 metres. It was speculated that these fields might be the first in the USA to be developed with an FPSO. However, once again the presence of an extensive pipeline network nearby was decisive. The Na Kika platform connected a series of underwater wells with christmas trees installed, which in turn were connected to manifolds from where the oil was sent to the platform through flexible risers. In contrast, by the early 1990s, the use of automated christmas trees and manifolds at extreme depths was now becoming a well-proven technology. However, like FPSOs the platform had no drilling capacity.

BP built on its experiences from Na Kipa when it ordered the world's largest semi-submersible production platform with drilling capacity for the Thunder Horse field. The platform, which was brought out to the field to be connected up in 2005, weighed 59,500 metric tons and was to operate at ocean depths over 1,800 metres. The platform was connected to a series of wells with wet christmas trees and corresponding manifolds on the seafloor. It had drilling equipment, making it possible both to drill new wells in the existing field and to discover new, smaller fields nearby through horizontal drilling. Moreover, it had significant capacity to connect underwater fields nearby through so-called tie-back solutions. From this point of view, the Thunder Horse platform had all the capabilities developed thus far by modern offshore technology.

But with a price tag of five billion dollars, the Thunder Horse platform was expensive, and both BP and Thunder Horse were hit by the dramatic hurricane season in 2005. The platform was evacuated when Hurricane Dennis approached in early July. When the crew came back, the platform was leaning at a 20 degree angle and close to capsizing. The platform was righted again, and survived both Katarina and Rita, whose centre passed a bit further east shortly afterwards. However, in the review after the hurricanes, weaknesses were found in the construction, requiring significant repairs. The platform was thus not ready for production before 2008. At this point, BP had already started production from a third giant semi-submersible platform on the Atlantis field. The Atlantis platform did not have drilling capacity, but on the other hand it was operating a larger field at even greater depths (2,156 metres) than Thunder Horse. Just like Shell's development of TLPs, BP followed a common pattern for contracts and development strategies. Both Thunder Horse's and Atlantis' hulls were built in Okpo, South Korea, and the topsides modules were fabricated by Ray McDermott in Morgan City, Louisiana. Their subsequent integration took place at Ingleside, Texas (Offshore Technology, 2017). For both projects, the contract for underwater equipment went to FMC.

FPSO and learning from Brazil

In 2005, the number of FPSOs in existence or ordered globally was 138, but there was still not a single one close to what was considered the technological centre of the offshore industry. This was despite the existence of enthusiasts in the Houston oil community who were active spokespersons for using FPSOs on developments in the Gulf of Mexico (Lovie, 2017). As with a semi-submersible platform, an FPSO was connected to underwater wells and manifolds by flexible risers. However, they did not have drilling capacity. This meant that all drilling or further interventions in wells, either before or during the production process, had to be carried out by hired-in rigs. On the other hand, an FPSO had storage capacity and was thus not dependent on pipelines. Could access to pipeline networks alone explain why the USA appeared so different to other offshore regions?

It was certainly true that while it remained relatively cheap to connect to existing pipeline networks, the need for solutions involving large storage capacities was not urgent. But if a comparison is made of the use of FPSOs in Brazil and Norway with the many projects where FPSOs were not chosen in the US, it can be seen that the pipeline argument does not explain everything. First, there were also fields in the US that were located so far from existing pipelines that TLP, Spar and semi-submersible solutions all required significant investment in pipelines.

The use of FPSOs in Norway confirms that location in relation to existing pipeline networks is important. Most Norwegian FPSOs lie in relatively remote areas (Norne, Skarv, Goliat). However, Statoil also chose to use an FPSO in central parts of the North Sea with pipelines not very distant but where the field was relatively small (Glitne). But while Norwegian FPSOs are placed on medium or small fields, many of Brazil's much larger number of floating production ships are placed on major oilfields in deep water.

The difference between Norway and Brazil cannot be fully explained without taking into account historical and political conditions. If Norway saw an early and extensive development of pipelines, in particular for gas, this was because of the political requirement to ensure that the greatest possible proportion of value creation would take place nationally. Furthermore, right from the start there were strong political incentives to ensure that as few resources as possible were wasted, which in turn contributed to a relatively high rate of exploitation on each field. For this reason, there was pressure on developers to use large installations with drilling capacity, but which could also be connected to pump systems to inject water or gas into the fields in order to increase the rate of exploitation.

Had its political situation been different, Brazil might well have taken the time to develop an infrastructure that could ensure a higher rate of exploitation. The oil reserves in the Campos basin were so large that a long-term development strategy for the whole area would probably have made large production platforms with drilling capacity, together with building several pipelines, more economic. But in Brazil, where the state oil company Petrobrás was never fully financed over the state budget, the company had strong incentives to choose solutions that would give the quickest possible extraction. The use of FPSOs was well suited to early production solutions that could contribute to financing further gradual development.

In the US, the oil industry was also under pressure to ensure a good long-term extraction plan, if not to the same degree then at least in the same way as in Norway. The difference here was that the pressure came from the interests of private capital, not from political signals as in Norway and Brazil (even if these signals went in very different directions). This contributed to the fact that, in the Gulf of Mexico too, the development solutions that were most commonly chosen had continuous drilling capacity and pipelines to land, even in fields smaller than those developed with FPSOs in Brazil. What is noteworthy is that despite this, there were not (as there were in Norway) more exceptions where it nonetheless made economic sense to use storage capacity and ship oil onshore from the fields using tankers. The Gulf of Mexico was large enough that many of the deep-water fields were dozens of kilometres or more away from the nearest pipeline. Laying pipes at 1,000–2,000 metres depth could be both expensive and difficult. Moreover, the underwater terrain was often both hilly and very steep (Minerals Management Service, 2001).

The advocates of FPSOs as a development concept pointed out that there were economic disadvantages of being linked to a specific pipeline. In contrast to Norway, where most of the pipeline network was either state-owned or where access was strongly regulated, pipeline ownership, operatorship and regulation in the Gulf of Mexico were more complex. By connecting to a particular pipeline, an oil company could find itself somewhat dependent on the refinery at the other end. By storing oil in FPSOs and shipping it to market in shuttle tankers, an oil company would be more flexible and could thus deliver wherever prices were highest at any given time.

While the shuttle tankers which received crude oil from an FPSO in Norway or most other places in the world could sail to any other country they liked, it had been prohibited to export oil from the US since the 1970s energy crisis (The Energy Policy and Conservation Act of 1975). This was tied up with significant conflicting interests inside the oil industry. Oil companies which made most of their income from selling crude oil pushed to lift the export ban. Independent refineries stood in a stronger negotiating position vis-á-vis the oil producers, and wished to uphold the ban (The Economist, 2015). Under the Obama presidency, a growing environmental movement also opposed lifting the ban. The argument was that this would be an inducement to increase production further, which in the last instance meant greater climate emissions.

But the strong growth in US production of shale oil, together with the low oil prices from 2014, changed most of the previous parameters. The US was suddenly no longer dependent on imported oil. The lifting of the ban in December 2015 was a political compromise where many Democrats voted for the measure together with Republicans in return for Republicans not blocking a vote of 500 million dollars to the UN's climate fund, and a measure on tax cuts for solar and wind power. The effect of this compromise, which was thus about far more than the offshore situation, was the end of a regime which in the final analysis set constraints on the kind of technological solutions that were economical for developing large oilfields.

Furthermore, the use of shuttle tankers in particular was another partial sector of the offshore market where American regulations had significant impact. Following the enactment of the Merchant Marine Act, the so-called Jones Act from 1920, both shipping and shipbuilding in the US had a strong protectionist character. Based on this law, all shipping between American harbours took place on US-owned and -built ships (Pub. L. No. 66–261, 41 Stat. 988, 1920). This meant that all the tankers which could ship oil from FPSOs to land would have to be built in the USA.

The Jones Act, and the question of which parts of the American regulation regime should regulate what, contributed to tensions in the offshore branch in the Gulf of Mexico from the 1990s and well after the turn of the century. Interest groups tied to the shipbuilding industry in the coastal regions along the Gulf of Mexico wanted to see TLPs, Spar platforms and semi-submersible platforms subjected to the maritime shipping regime and thus the Jones Act, in the same way as supply ships and floating exploration rigs. If these groups had won, neither the hulls of Shell's TLPs nor those of the many Spar platforms or BP's semi-submersible Thunder Horse, could have been built abroad.

However, the final legal clarification did not come until after a series of court cases well into the 21st century (*Case* v. *Omega Natchiq* 2008. v. *Anadarko*, 2012; *Hefren* v. *Murphy* 2014). The oil industry succeeded in defining FPSOs as lying outside the Jones Act. This was clearly a grey area since FPSOs were vessels with a clearly maritime character. Furthermore, many FPSOs abroad used dynamic positioning rather than anchors to hold themselves on the right point on the oilfield. But as long as they produced oil, they had a connection to the ocean floor through flexible risers; at least, this was the interpretation which ruled that the Jones Act did not apply. By contrast, the shuttle tankers which would have had to ship oil to and from FPSOs in the Gulf of Mexico could hardly be defined as falling outside the Act. Insofar as American tankers, built in the USA, with American sailors, were far more expensive than the situation in most other offshore regions, this constituted a cost-increasing condition which benefitted Spar, TLP and semi-submersible solutions at the expense of FPSOs.

Deepwater Horizon changes the parameters

While the American regulation regime around shipping contributed to excluding FPSOs through expensive regulations, there was also widespread scepticism about the use of FPSOs in the Minerals Management Service, which regulated production installations. After the Exxon Valdez disaster in 1989, there was general scepticism about extensive storage of large amounts of oil at sea in the USA. Following the large compensation payments Exxon was forced to make, this was a parameter that the oil companies also had to bear in mind when selecting technology. Here, it is worth noting an important difference to Norway. In the 1970s, there was a widespread fear in Norway that large levels of oil pollution would affect the coast. When an Ekofisk field was hit by a major blowout in 1977, large quantities of oil leaked into the sea. The combination of the field's location, far out to sea, and a storm that broke up the oil long before it reached land, lessened concern about oil emissions. Although in the years that followed, Norway would build several pipelines, a considerable proportion of oil would be transported via offshore loading systems out on the fields. The same was true for Great Britain. This contributed to the fact that in 1996 the North Sea had a fleet totalling 31 shuttle tankers in operation, while there were none at all in the US Gulf of Mexico.

When, in the late 1990s, Texaco became the first company in the US to seriously consider an FPSO as a development process, the MMS launched a review process (Upstream, 1998). The upshot was that FPSOs could be used, but with particularly strict safety requirements because of the risk of pollution. By the time the review was finished, however, Chevron had taken over Texaco and chose another solution. It might have been expected that the many accidents associated with TLPs and semi-submersible installations around the 2005 hurricane season would be an argument in favour of FPSOs, but the conclusion was generally the opposite. There was speculation as to what might have happened if the many installations that were floating around out of control had collided either with FPSOs or associated shuttle tankers filled with oil.

With all these particular, often socially-determined conditions, it is then no accident that it was the Brazilian state oil company Petrobrás which became the first to use an FPSO for a deep-water development in the US. The decision to use an FPSO in the US was already taken before the Deepwater Horizon accident in 2010. This accident led to far greater emissions than could have been expected even from a fully-loaded tanker. There was thus a change in favour of the use of FPSOs from the point of view of risk assessment. From 2012, the vessel BW Pioneer produced oil and gas from the Cascade and Chinook fields. Petrobrás had bought into a refinery in the Houston area in 2008; when full production started from BW Pioneer. However, the company used its freedom to sell oil where the prices were best, in the first instance within the US. But from December 2015, the company was free to sell its oil outside the US.

Furthermore, the first FPSOs in the USA represented a shift in the contract model which had previously been used in the Gulf of Mexico. BW Pioneer was both owned and operated by the Norwegian FPSO shipping company BW Offshore (Offshore Technology, 2017). Like many projects on the Brazilian continental shelf, BW Pioneer was a modified oil tanker. The remodelling contract went to the Keppel Corporation at its Singapore shipyard. The project was thus distinct from other major developments in that much of the processing equipment was installed outside the US (Keppel, 2008). But just as with simultaneous developments in the Norwegian sector, established offshore companies took care of the engineering work and supplies of central equipment. The topside engineering contract went to KBR, which had absorbed the traditional offshore supplier Brown & Root.

In 2013, Shell decided to use an FPSO as a production concept on the Stones field. Because the field was very deep (2,896 metres) and smaller than many earlier deep-water developments, it was not appropriate to use large, fixed platforms. Moreover, the development was one of the first projects where the decisive choices were made subsequent to the Deepwater Horizon accident. The dramatic hurricane season in 2005 was still foremost in people's minds when the development concept was decided upon. Turritella, the name of the FPSO that was used, featured a turret assembly with a buoyant mooring that allowed the FPSO to rotate freely around the buoy as the wind and currents change. The mooring system is permanently fixed to the seabed while the FPSO can disconnect from the buoy if a hurricane approaches (Oil & Gas Journal, 2017). This innovation made the use of FPSOs far more attractive for future developments of small and medium fields in the Gulf of Mexico. Once again, we see how social conditions affect the choice of designs, which in turn advances the development of particular technologies.

Concluding remarks and consequences for Norwegian suppliers

This overview of major field developments in the American sector of the Gulf of Mexico, seen in the light of comparable projects in Brazil and the North Sea, shows that it has not been only economic responses to geographical and geological conditions, but largely also to social and political conditions which have been decisive for the development of offshore technology. Furthermore, the historical, social and political preconditions which contributed to shaping technological choices offshore in the US help to make clear how other, similar, social conditions have affected developments in other offshore markets. These preconditions were also important for many Norwegian suppliers, which from the early 1990s and on worked hard to succeed in the US Offshore market (Chapter 9).

Even though the offshore oil market in the Gulf of Mexico was dominated in the 1990s by a more cohesive group of oil companies with operative responsibilities, these appeared less oriented towards consciously creating competition between the main EPC suppliers than Norway, Brazil or Great Britain. This led to a few individual supply companies having a near-monopoly position for long periods as regards deliveries of some key technologies. For Aker Solutions, a company that for years tried to establish itself as a EPC supplier of subsea equipment, the barriers to entry was high. Like other Norwegian companies it got most of it contracts in the role as sub-supplier.

In all the main markets for offshore contracts we have discussed, there has been different types of protectionism. In the US, the Jones Act and the ban on oil exports in particular have had a great impact on both technological choices and market access. Protectionist regulations have generally encouraged the use of large installations without storage capacity, rather than more maritime solutions based on using FPSOs. This was generally not an advantage for Norwegian companies, which had one of its strength in the maritime part of the offshore market. The Jones Act resulted in Norwegian supply ship owner with some few exceptions was shut out of the American market. This, while the same part of the Norwegian offshore industry got a large market share from Brazil (Chapter 10)

The rate of exploitation from an oilfield depends on a combination of reservoir quality and technological choices. By comparison with Brazil, where technological choices were affected by the lack of investment and pressure to secure income quickly, the extended use in the US of development solutions with fixed drilling capacity, tied to refineries on land by pipelines, led to a relatively high rate of exploitation per field. In this area, however, the Norwegian regulation regime has distinguished itself by furthering a long-term approach and with a strong emphasis on securing the greatest possible rate of exploitation.

Just as the blowout on Ekofisk in 1977 and the capsizing of the Alexander Kielland in 1980 left their mark on technological developments on the Norwegian continental shelf, so too the Exxon Valdez accident in 1989 had a long-term and major impact on offshore technology choices in the US. After the accident, there was a long-lasting and widespread opposition to technological solutions which involved large stores of oil at a single location at sea. This strengthened an already-present aversion to the use of FPSOs combined with shuttle tankers, both a market where Norwegian companies were strong. However, the dramatic hurricane season in 2005 and the Deepwater Horizon accident in 2010 have contributed to change risk perceptions in a way which, unlike previously, encourages the use of FPSOs. It was no accident that the first FPSO operating in GoM was operated and owned by a Norwegian company.

References

Engen, O. A. and Olsen, O. E. (1997). Konservativ nyskaping i Offshore Oljeproduksjon. Olsen, O. E. and Sejersted, F. (eds), *Oljen som teknologiutviklingsprosjekt*. Oslo: Gyldendal.

Odemis, B. B. (2015). *The Nature of the Firm in the Oil Industry: International Oil Industry*. Abingdon: Routledge.

Parmentiier R. (1999). Greenpeace and the dumping of waste at sea: A case of non-state actors' intervention in internatioal affairs. *International Negotiation*, 4(3).

Ryggvik, H. (2013). *Building a skilled national offshore oil industry. The Norwegian experience*. Oslo: Confederation of Norwegian Enterprise (NHO).

Additional sources

Case v. *Omega Natchiq*. (2008), Inc., 2008 WL 2714124 (S.D. Tex. 7/10/08).

Enze, C. R., Brasted, L. K., Arnold, P., Smith, J. S., Breaux, J. N. and Luyties. (05.1994 W. H. Shell Auger TLP Design, Fabrication, and Installation Overview. Presented at the 26th Annual OTC in Houston, Texas, 2–5. May 1994.

Hefren v. *Murphy Exploration and Production Company*. (2014).

Houston Chronicle. (29.09.2005). Storm left big platform upside down.

Keppel Corporation. (1.02.2008). Keppel wins FPSO conversions of over S$215 million from repeat customers.

Lovie P.M. (2017). FPSP in GoM – a 21 Year Saga. PPP 2017.

McDermott International. (2012). Annual report 2002.

Mendez v. *Anadarko*. (2012). 466 Fed.Appx. 316 (5th Cir. 2012).

MMS. (2001). Brief Overview of Gulf of Mexico OCS Oil and Gas Pipelines: Installation, Potential Impacts, and Mitigation Measures, ICS Report MMS 2001–067.

New York Times. (24.04.2006). Shell to reopen platform in the gulf.

Oceanstaroec. Hall of fame. Retrieved 23.02.2018 www.oceanstaroec.com.

Offshore Magazine. (11.01.2000). Why Shell chose semi over TLP for Na Kipa.

Offshore Magazine. (15.01.2013). Finland yard maintains leadership in GoM Spar hulls.

Offshore-technology. Atlantis Deepwater Oil and Gas Platform, Gulf of Mexico. Retrieved 23.02.2018 www.offshore-technology.com/projects/atlantisplatform/.

Offshore-technology. Cascade and Chinook Subsea Development, Gulf of Mexico. Retrieved 23.02.2018 www.offshore-technology.com/projects/cascadechinook/.

Offshore-technology. Ram Powell Oil and Gas Field Project. Retrieved 2017 www.offshore-technology.com/projects/rampowell/.

Oil & Gas Eurasia. (3.12.2008). Shell Drills World's Deepest Offshore Well.

Oil & Gas Journal. (1.07.1996.) Chevron plans spar installation in U.S. gulf.

Oil & Gas Journal. (23.09.1996.) Launch imminent of world's first production spar system.

Oil & Gas Journal. (5.01.2017.) Shell's Stones field advances subsea production technology.

Petroleumstilsynet. (28.11.2004). Gransking av gassutblåsing på Snorre A Brønn 34/7-P31.

Pub. L. No. 66–261, 41 Stat. 988. (1920). Merchant Marine Act of 1920.

Shell. (6.09.2016). Shell starts production at Stones in the Gulf of Mexico. (Press release).

The Economist. (28.03.2015). Retrieved 21.12.2015. Binning the ban. A ban that benefits only American oil refiners is under fire.

Upstream. (30.05.1998). Texaco takes FPSO plunge. US major focuses on Fuji for Gulf of Mexico breakthrough.

Wood Group Mustang (2015). Worldwide Locations of Deepwater Facilities and Status as of March 2015.

Wood Group Mustang. (2012). Worldwide survey of Spar, DDCV, and Mini DOC Vessels.

10 Steel, staff and solutions

Past, present and future prospects for employment in the Norwegian-based petroleum supply industry

Atle Blomgren and Christian Quale

Introduction

This chapter analyses the growth, current structure and possible future of employment in the Norwegian-based petroleum supply industry. From its start in 1965, this industry has grown to around 160,000 employees (Blomgren *et al.*, 2015) of which 50 per cent in foreign-owned companies (Sasson and Blomgren, 2011). To analyse the structure of employment, we will divide industry employment into three generic functions:

- Headquarters functions for rigs and vessels, for use in both the home and export markets ('Steel').
- Development and provision of products and services for the home market ('Staff').
- Development and provision of products and services for export markets ('Solutions').

We start by presenting how Norwegian nationals and Norwegian-owned companies gradually 'mastered the trade'. We then analyse current employment divided into functions, geography and industry sector. Finally, we use the preceding analysis to speculate about the future prospects of employment in the Norwegian-based petroleum supply industry and the possibilities of gradually more non-petroleum related employment.

The main sources of data for this chapter are two surveys for the Norwegian Oil and Gas Association (Blomgren *et al.*, 2015, 2013). These two surveys contain workplace/plant-level data for all plants in the Norwegian-based petroleum supply industry and 170 historical case studies. The workplace/plant-level data includes all officially-registered data plus non-registered data gathered from media reports, websites and/or surveys of the various companies: share of activity at plant related to oil and gas, share of exports at plant related to oil and gas, etc. We also use data from some additional interviews with industry insiders.

How Norwegian companies 'mastered the trade' and became central actors in the global petroleum industry

In an article about the birth of the Norwegian petroleum era, *The Economist* wrote: [T]he first oil was extracted … in 1971 … turning Bergen from a fishing village into an industrial hub.' (*The Economist*, 2015). A few days later, the following comment appeared on the Economist's website:

> It was not Bergen that became an industrial hub, but Stavanger. And to call it a fisheries village is a vulgarity. Stavanger had fisheries related industry and shipping traditions, both of which were strongly globalised, long before the petroleum age.
>
> (*Economist* comments, 2015)

The anonymous commenter is right in asserting that when the petroleum industry came to Norway, it did not enter into a void, but into a society capable of taking an active part in its development. Industry and institutions (Hunter, 2014) collaborated in developing a strong home market-based industry ('Staff') with technological capabilities for exports ('Solutions'), even though the home market focus at times ran counter to the interests of ship and rig owners ('Steel').

The Norwegian-based petroleum supply industry dates back to 1965 with the establishment of supply bases to serve the drilling rigs that were expected to arrive from the US the following year. North Sea Exploration Services AS was established by the shipping companies Staubo (Oslo), Grieg (Bergen) and Smedvig (Stavanger), while Norwegian Oil Repair & Supply Co ('Norsco') was established by the shipping company Fearnley & Eger and the largest Norwegian industrial company, Akers Mek. Verksted (Blomgren *et al.*, 2015; Pahr-Iversen, 2006). In 1966, Aker's yard in Oslo started the construction of its first drilling rig, and the classification society Det Norske Veritas (today part of Det Norske Veritas – DNVGL) started developing a special class for offshore drilling rigs (Hanisch and Nerheim, 1992). Also in 1966, Norwegian shipping companies established the two companies that would eventually develop into the world's two largest drilling rig companies: Aker Drilling Company Ltd. (later Transocean) and Smedvig Offshore (later Seadrill). In addition to competence in handling vessels in tough offshore environments, the Norwegian shipowners brought with them financial expertise in ordering vessels on speculation without having a firm drilling contract, a practice that hitherto had not employed in offshore drilling (Hanisch and Nerheim, 1992).

The drilling rig company, Ocean Drilling & Exploration Co. (Odeco), and the service company, Schlumberger, both established themselves in Norway (Stavanger) in 1966 (Gjerde, 2012).[1] The companies promptly started hiring Norwegian nationals for 'non-oil' positions. The very first job advertisement by Odeco explicitly looked for 'skilled and unskilled personnel' and mentioned in particular 'electricians, welders, motormen, labourers and second class radiomen'

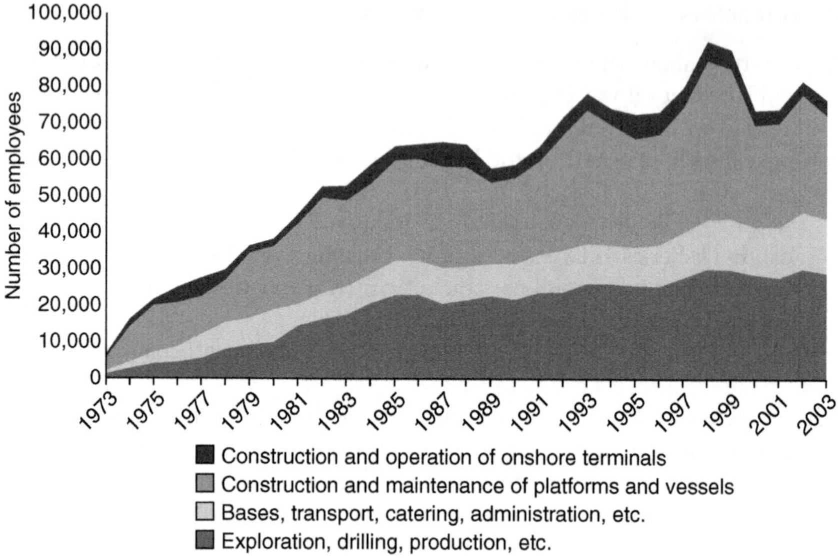

Figure 10.1 Petroleum-related employment (incl. oil and gas companies) in Norway, 1973–2003.

Sources: IRIS and The Norwegian Labour Market Authority.

(Gjerde, 2002, 14), preferably with a working knowledge of English (Krogh, 2002). By the mid-70s around 90 per cent of Norwegian nationals working on drilling rigs were former fishermen or seafarers (Tellnes *et al.*, 2003). As Norwegian nationals with drilling expertise were not available, Norwegian shipowners engaged in drilling rigs had to participate in joint ventures and/or hire foreign nationals (Hanisch and Nerheim, 1992).

Sensing opportunities for local businesses, the local authorities in the Stavanger area were quick to provide for the petroleum industry, providing everything from housing to English language schooling to golf courses (Gjerde, 2002). After the discovery of the giant Ekofisk field in late 1969, central Norwegian authorities started taking an active role in the build-up of the industry (Lie, 2012). Norwegian universities were given resources to focus on petroleum specific disciplines (geology, geophysics, drilling and reservoirs) that were more or less absent in the Norwegian educational system (Hanisch and Nerheim, 1992). In 1972, the Stavanger Chief Engineer School (today: Stavanger Offshore Maritime School) commenced a seven week course in drilling, and the first one-year program in 1974 (Tellnes *et al.*, 2003). In general, Norway's system of vocational training and apprenticeships ensured a steady flow of well-trained skilled labour quite capable of interacting directly with the higher-educated personnel. An informant from one of our surveys argues that the resulting skill level of 'ordinary' workers – 'skilled labour' – made possible fruitful collaborations between engineering and operations: '*Our R&D emerges from the close collaboration between the engineers in our development department and the skilled labour in our operations department.*' (Location manager for large equipment supplier).

To ensure local content, the Norwegian government enacted a number of 'infant industry'/'local content' policies (Engen, 2009; Hunter, 2014) (see also Chapters 2 and 7). First, the Royal Decree of December 8th, 1972 stated that Norwegian commodities and service should be used when competitive. Even though the decree was not used strictly for protectionism, the Norwegian shipping community was opposed to it as they feared other petroleum provinces would 'retaliate' (Ryggvik, 2013). Second, the Norwegian government used the system of concessions for operations on the Norwegian continental shelf (NCS) to ensure that foreign oil companies funded Norwegian R&D and collaborated with Norwegian universities and research institutes (Ryggvik, 2013). Third, the state ensured that important state-owned companies became involved in the industry, and in 1973 the two state-owned companies, Statoil and Kongsberg Våpenfabrikk, established the reservoir analysis company Statex (which in 1977 merged with a spin-off from DNV to create Geophysical Company of Norway – Geco). Fourth, the government established dedicated programs for financing petroleum-related R&D and promoting collaboration between the government, academia, operators and suppliers (see Chapters 2 and 3).

As the development of the Norwegian oil and gas resources presented the oil companies with a number of hitherto unknown technological challenges, R&D collaboration in Norway resulted in a wide range of new technological *solutions* that were subsequently used in most other offshore provinces (see also Chapters 2 and 6). These included, among others: 1. Dynamic Positioning by NTH/Kongsberg Maritime (Bjørnstad, 2009); 2. Multilateral wells by Halliburton (Grønås and Hundsnes, 2002); 3. Accurate horizontal drilling by Baker Hughes (Norwegian Academy of Technological Sciences, 2005, 39); 4. 3D and 4D seismic by Geco/Schlumberger[2]; 5. Remote-controlled, diver-less module-based subsea installations by NUTEC/SINTEF/Kongsberg (Bjørnstad, 2009; Rystad Energy, 2013b, 51); 6. Multi-phase flow measurement by CMR/Fluenta (Norwegian Academy of Technological Sciences, 2005, 75); 7. Multi-phase transport by SINTEF/IFE/Scandpower (Norwegian Academy of Technological Sciences, 2005), etc.

However, not all technologies developed *for* the Norwegian continental shelf were actually developed and produced *in* Norway. To be able to produce the thin oil layer in the Troll oil province, Halliburton and Baker Hughes developed brand new technologies for multilateral wells and accurate horizontal drilling. While the Norwegian subsidiaries of Halliburton and Baker Hughes participated in the development, and there was extensive testing in Norway, the bulk of development and manufacturing took place outside the country, and neither of these technologies became export successes for the Norwegian-based supply industry.

Aside from R&D, another important factor for the global take-up of Norwegian-developed technologies was the eventual global acceptance of the relatively strict Norwegian safety standards. The collapse of the Norwegian accommodation platform, Alexander Kielland, in 1980 with 123 lives lost, gave new impetus to improving offshore safety. In 1981, Rogaland Research started a project on improved safety through automation of drilling operations inspired by the focus on separating man and machine at the Institute for Energy Technology's (IFE) nuclear reactor at Halden. The project led to automation technologies that were commercialised by Jon Gjedebo's Hitec (Bøe, 2012; Norwegian Academy of Technological Sciences, 2005). One of our informants points out that in some segments strict Norwegian regulations functioned like a barrier to entry that sheltered Norwegian companies from global competition, and that the relative importance of strict Norwegian regulations may explain differences in export success:

> The development of topside drilling equipment was driven by Norwegian regulations, and the companies of Gjedebo and Skeie [the two entrepreneurs behind what is today NOV's Norwegian activities] were able to grow within what was in effect a closed market. The market for down-hole tools is different, it is not driven by Norwegian regulations and one has to compete globally from day one.
>
> (Entrepreneur behind down-hole well technology company)

The Norwegian state holds ownership stakes in two important supply companies (the Kongsberg Group and Aker Kværner Holding), but the market for ownership in the supply industry has generally not been restricted.[3] Both Norwegian and foreign companies have been acquiring companies in order to access skills or gain a foothold in new markets. To gain needed expertise in drilling, the US platform drilling company, Morco Norge, was taken over by Norwegian-owned Norcem/Aker and used for the development of what would become Transocean and Prosafe (Prosafe NOW, 2002). The Norwegian companies, Stolt Offshore/Acergy (Haugesund) and DSND Subsea/Ugland (Grimstad), were merged and in turn acquired French Comex and US Halliburton Subsea to create the subsea giant Subsea 7 (Ryggvik, 2013).[4] NorSea Group, the result of a merger between the first two Norwegian supply bases, has expanded abroad by acquiring supply-base companies in the UK and Denmark.

There are also numerous cases of successful Norwegian-owned companies being taken over by larger, foreign companies (see Chapter 6 and Ryggvik (2013), and a survey from 2008 found 50 per cent of Norwegian-based employees were working in foreign-owned companies (Sasson and Blomgren, 2011). In 1993, the ship design and maritime equipment division of the Ulstein Group was acquired by British Vickers and subsequently sold on to British Rolls-Royce, for which it became a new corporate division, Rolls-Royce Marine (Blomgren *et al.*, 2013). In 1990, US/French Schlumberger secured 100 per cent control of Geco, and its activities were partly incorporated into the marine seismic company, WesternGeco, and partly into Schlumberger's division for Exploration and Production (E&P) software (Blomgren *et al.*, 2015). Being acquired by a larger corporation provides access to both capital and global distribution channels. The following quote from a book on the history of Geco indicates that the take-over by a larger actor like Schlumberger was necessary to safeguard employment and make possible further international expansion:

> By the summer of 1986, Geco had the choice of either being reduced to a minor marine seismic company with a handful of vessels, or to join a large international company with capital and technology to further develop its activities.... It was an easy choice for both owners and employees.
>
> (Bjerke, 1989, p. 101. *Authors' translation*)

In addition to pure takeovers, there are also numerous examples of Norwegian–foreign joint ventures (Aibel, Island Offshore, etc.) and companies owned by Norwegian private equity funds funded in part foreign capital (Apply Sørco, Ocean Installer, etc.).

All cases mentioned above are examples of foreign capital boosting national employment by providing capital for local expansions and/or global distribution channels. One of our informants point out that foreign ownership, in particular foreign ownership that agrees to keep activities in Norway, is a key to global success in the Norwegian-based industry.

> Foreign ownership is not a bad thing, NODE [the lifting and drilling equipment cluster in southern Norway] did not take off before National Oilwell bought Hydralift. The key to export success from Norway is access to distribution channels and a foreign owner that accepts development and manufacturing in Norway.
>
> (Owner of large petroleum supplier)

However, not all foreign takeovers result in boosting national employment. The headquarters functions of both Transocean and Smedvig were gradually moved out of the country after foreign takeovers in 1996 and 2005 (Ryggvik, 2013). EasyWell developed the so-called Swell Packer technology into an export success, but after Halliburton took over in 2005 all development and manufacturing was gradually moved out.

Much of the revenue from the foreign takeovers has been used to start new businesses, so-called serial entrepreneurship. The revenues from the sale of EasyWell were reinvested in the well technology company, Fishbones. The capital gains from the Ulstein family's sale to Vickers were used to establish the Ulstein Group within ship building, design and equipment, and Island Offshore within offshore shipping (Blomgren *et al.*, 2015, 2013). The profit from the sale of Hitec to National Oilwell was used to establish Norway's largest private equity fund, HitecVision.[5]

From handling provisions to rigs and working as mere roustabouts in 1966, the last 50 years have seen both Norwegian-based companies and Norwegian nationals becoming central players in the global petroleum supply industry.

Several Norwegian nationals have held global positions, and at present the most prominent is Paal Kibsgaard who, since August 2011, is the global CEO of the world's largest petroleum supply company, Schlumberger. To prove that Norwegian nationals really did master the trade, one of our informants point to the large number of Norwegian nationals in the global hierarchies of the petroleum industry.

The service companies hired Norwegians, and gradually the locals mastered the trade. Today you will find Norwegian nationals in senior positions in most global drilling and well service companies (Former CEO of one of the largest drilling and well service companies in Norway). In this section we have shown how Norwegian companies and Norwegian nationals gradually 'mastered the trade' of the petroleum industry. In the next section we will study how this success affected Norwegian-based employment in the petroleum supply industry. We will focus on both total Norwegian-based employment and employment divided into three different functions (activities towards home market, activities towards export markets and activities within headquarters functions).

Current employment in the Norwegian-based petroleum supply industry

In this section we present workplace/plant-level data on the 160,000 employees in the Norwegian-based petroleum supply industry (Blomgren *et al.*, 2015, 2013) and analyse drivers for employment.[6] The employment data is from the end of 2014/early 2015, which is generally considered the 'peak' years in this industry. Plant-level data makes it possible to distinguish between different functions within companies. A given company may consist of three units/plants, a factory producing equipment for exports, a service unit for the Norwegian continental shelf and a unit hosting headquarters functions. In this example the first unit will be export related, the second will be home market related while the third is headquarters functions for both the home and export markets.

There is, however, an inconsistency in the official registers regarding employees offshore as employees on fixed and movable offshore installations are assigned with an *offshore* workplace/plant, while seafarers are assigned with workplace at the shipowner's *onshore* address. Shipowners tend to distribute their employees over a number of subunits, so we identified their onshore employment by singling out subunits with relatively few employees and/or enterprise names containing the term 'management'; the remaining subunits were assigned with workplace offshore. For rig- and shipowners we thus have a distinction between operations (offshore) and headquarters functions (onshore).

In Figure 10.2, the chart to the left shows employees in directly petroleum-related Norwegian-based companies by generic function and region (by county – larger administrative regions in Norway). We see that there is petroleum-related employment in every county, but that the bulk of home market activity is in the 'offshore' county and in the two counties geographically closest to the operations on the NCS, the south-western counties of Rogaland and Hordaland.

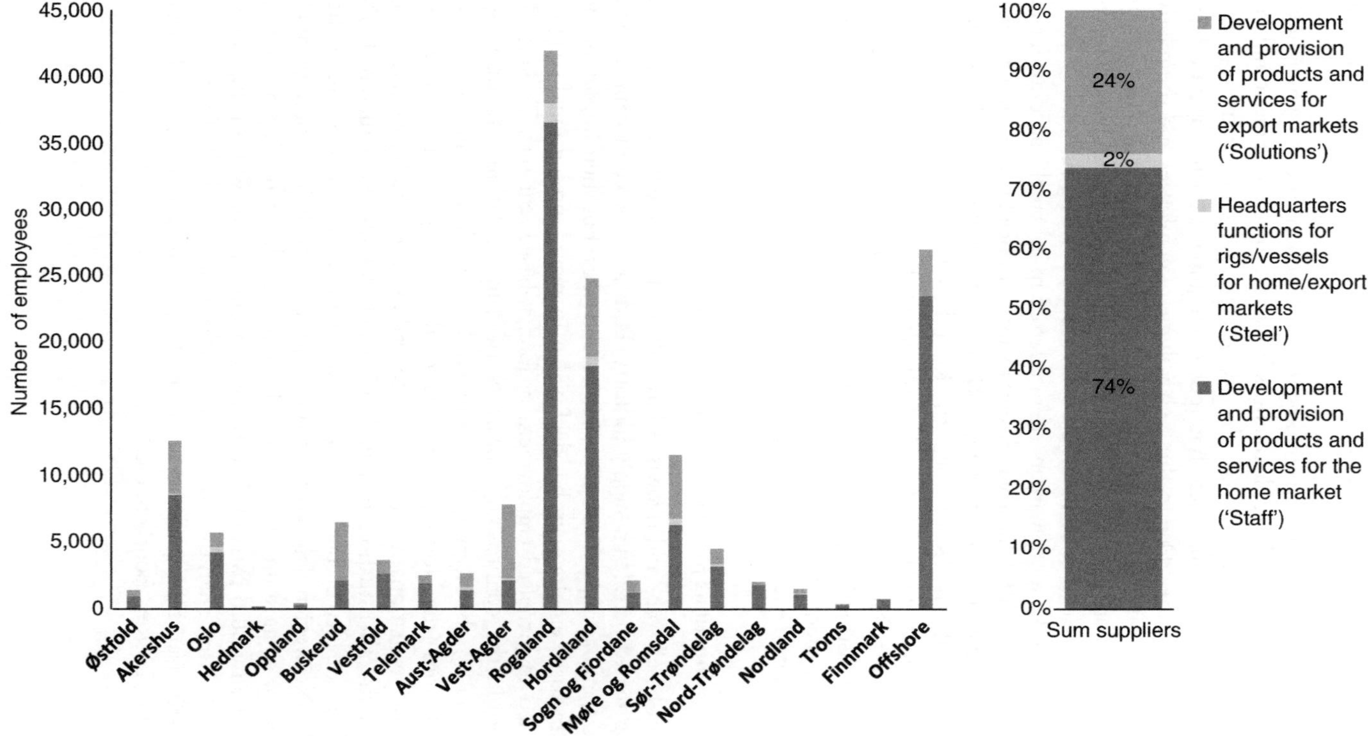

Figure 10.2 Employees in Norwegian-based businesses directly related to petroleum activity by county and generic function, 2014.

Source: based on Blomgren *et al.* (2015).

This indicates the importance of proximity to operations; it is obviously easier to operate repair workshops close to petroleum activities than across the mountains in eastern Norway.

The right-hand chart sums up employment by function. We see that approximately three-quarters of all employment is linked to the development and provision of products and services for the NCS. Here, we find a combination of equipment suppliers, companies that rent out equipment with operators and pure staffing companies. Due to the high volume of employees and the relative high labour intensity we will refer to this as 'Staff'. Export-related activity (referred to as 'Solutions') employs slightly less than a quarter of all employees. Headquarters functions of companies owning and managing rigs and vessels in both the home and export markets (here referred to as 'Steel') account for per cent of employees. Note that there is employment in headquarters functions in other parts of the industry as well, both in large groups administered *from* Norway (DNV-GL, the IKM Group, the Kongsberg Group, etc.) and in global headquarters functions located *in* Norway (Rolls-Royce's marine division is administered from Norway). But the case of rig- and shipowners is unique in having a clear distinction between employment in operations (offshore employment) and in headquarters functions (onshore employment). Below we discuss the employment in each of these three functions in more detail.

Headquarter functions for rigs and vessels ('Steel')

Norwegian-based ship- and rigowners have always had a crucial position in the Norwegian-based petroleum supply industry, both as a source of competencies and capital, and as important end-users of technology. But their importance for Norwegian-based employment has declined as increased international competition has led to increased employment of foreign-based seafarers. To keep the shipowners in the country and ensure Norwegian regulations on the vessels (Norwegian flag), the Norwegian government has introduced a number of beneficial regulations (the Norwegian International Ship Register, tax exemptions, etc.). These regulations do not extend to drilling rigs which are categorised together with 'other offshore installations' (fixed and movable structures). The result of this duality is that while many *shipowners* keep a large share of their fleet under Norwegian flag and keep the major parts of their onshore organisations in Norway, most *drilling rig owners* have moved the ownership of their rigs abroad and often also parts of their onshore organisations. Within shipowners, we find both Norwegian-owned companies (Solstad, Farstad, DOF, etc.), joint ventures (Knutsen NYK Offshore Tankers, Island Offshore, etc.) and foreign-owned firms (TeeKay, Bourbon, etc.).

Development and provision of products and services for exports ('Solutions')

Even though export-related activity represents only about a quarter of total supply industry employment, the absolute number of employees is substantial. By 2014, the Norwegian petroleum supply industry was regarded as one of Norway's largest export industries after crude oil and natural gas (Nordbø and Steinsland, 2015). The explosive growth in exports from the turn of the millennium (see Figure 10.3) is mainly attributed to the success of equipment suppliers within drilling and handling equipment (NOV, MHWirth, McGregor), subsea (FMC, Aker Solutions, OneSubsea) and maritime equipment (Rolls-Royce, Kongsberg, Wärtsilä, Vard, Ulstein) (See Chapter 7).

The fact that most of the important exporters are foreign-owned companies (Rystad Energy, 2017) begs the question why a foreign owner would want to conduct export-related activities from a high-cost country like Norway. First, industry insiders indicate that the costs of *higher-educated personnel* in Norway are reasonable in an international setting. Second, as each field development is unique, the oil and gas sector has little place for serial manufacturing and thus may accept relatively more high-cost, customised solutions (Beyazay-Odemis, 2016). Third, the exporters *have* taken steps to reduce costs wherever possible, and much fabrication is now sourced from abroad. Fourth, some of the exports are used as inputs for, and installed on installations destined for use on the NCS (Hungnes *et al.*, 2013), suggesting some kind of 'home field advantage' for Norwegian-based exporters. Fifth, the Norwegian industry clusters are said to provide opportunities for collaboration that is hard to replicate elsewhere; see for instance the complete maritime cluster in Møre og Romsdal (Aamodt and Frøshaug, 2012), the drilling and handling equipment cluster in southern

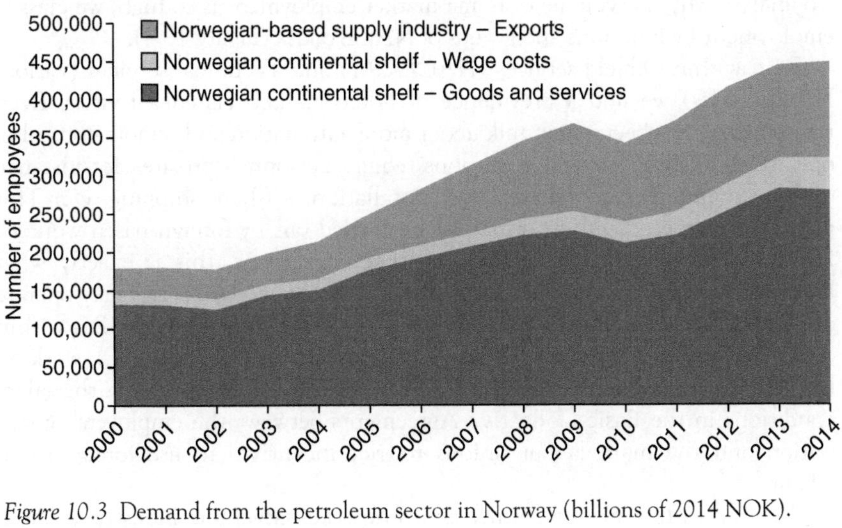

Figure 10.3 Demand from the petroleum sector in Norway (billions of 2014 NOK).

Source: Nordbø and Steinsland (2015).

Norway (Isaksen and Karlsen, 2012) and the engineering/subsea cluster in eastern Norway (Ryggvik, 2013). Sixth, there are elements of path dependency as Norwegian-based companies have developed solutions that have become global standards (see Chapters 6 and 7), and there would be risks involved in moving the associated project organisations abroad. The entrepreneur of one of the largest export successes in the Norwegian supply industry sees little risk of business moving abroad as long as the company being acquired both has a series of unique products and a well-functioning project organisation, as the following quote illustrates:

> By the time we were acquired, [name of the company making the acquisition] simply did not provide the product categories that we offered. As we also had a well-functioning project organisation, they decided it was better not to move the business out of Norway, but to develop it as a central hub in the global organisation.
>
> (Entrepreneur and former owner of company sold to a major equipment supplier in the mid-2000s)

Development and provision of products and services for home market ('Staff')

Home market-related employment constitutes almost three-quarters of total employment. This helps explain the relatively low share of imports for investment on the NCS, around 40 per cent (Hungnes *et al.*, 2013), and indicates that Norwegian-based industry has managed to compete with foreign imports even after the most protectionist 'infant industry'-measures were abolished following Norway's entry into the EU internal market in 1994 (Ryggvik, 2013). To analyse why the volume of home market employment is so high, we classify employment by functions and groups of NACE codes (Figure 10.4).

First, within Oilfield services (NACE 09.1) and Technical services (various NACE codes) we find a prevalence of home market-related activities. Even though there has been much talk about more automation and remote control of operations offshore, several operations require personnel 'on-site', for instance offshore maintenance, well services, installation, offshore shipping, etc. This physical presence could, in principle, be carried out by foreign-based workers. For activities governed by global maritime regulations, this is exactly what happens; both marine seismic companies and accommodation rigs may operate on the NCS with foreign-based employees in all maritime positions. But within direct E&P-related activities (drilling, subsea work, well services, etc.), work on the NCS is carried out exclusively by Norwegian-based employees as the salary conditions in the Basic Collective Agreements between the employers' organisation and the main labour unions restrict incentives to use foreign-based labour.[7]

Second, within manufacturing industries, i.e. Machine industry (NACE 25–29) and Yards (NACE 30 + 33) we find significant employment related to

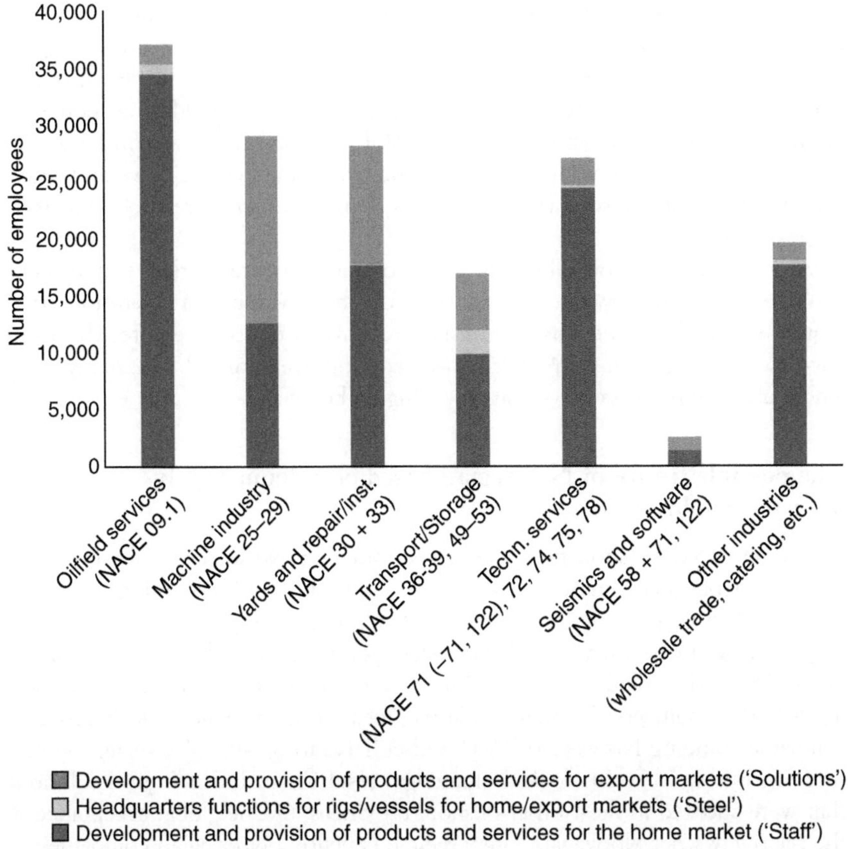

Figure 10.4 Employees in Norwegian-based businesses directly related to petroleum activity by generic and NACE codes, 2014.

Source: Based on Blomgren *et al.* (2015).

both home and export markets. 1. Here, we find activities that require physical presence that must be carried out using Norwegian-based workers, for instance maintenance and installation work offshore. 2. We find activities where physical presence to operations obviously is beneficial such as after-market services on equipment and construction activities with a high degree of customisation (Beyazay-Odemis, 2016). The large platform yards (Kværner, Aibel and Aker) seem to have a 'home field advantage' in customised solutions for the NCS and manage mostly to compete with foreign – mainly Asian – yards for installations on the NCS. An important sub-supplier in this segment is the IKM Group which has around 2.000 employees in Norway, and is as such the largest petroleum supplier 100 per cent owned by private, Norwegian interests. 3. Within Transport/Storage, home market employment is partly onshore employment within logistics/bases, and partly Norwegian-employed seafarers serving in

Norwegian waters (employees within drilling are which are included with Oilfield Services). The seafarers are to a large extent related to Platform Supply Vessels that are required to use Norwegian-based personnel as they go in transit between Norwegian ports, (a rig or an installation is defined as a 'port'). 4. Finally, within Seismic and Software (NACE 58 and 71,122) we find a 50/50-split between activities related to home- and export market. Here, we find activities within marine seismic (PGS, TGS, etc.) and E&P software (Schlumberger, Emerson, etc.).

As may be seen, the relatively high volume of home market employment could be explained by the following: 1. the high volume of operations that require physical presence and where political/union pressure requires this presence to be carried out by Norwegian-based employees, and 2. the many cases where physical presence to operations is highly beneficial.

The possible future of Norwegian-based petroleum supply industry employment

We commenced by showing some prerequisites for Norwegian companies and Norwegian nationals 'mastering the trade' and with time becoming a massive source of Norwegian-based employment: 1. National companies able to take an active part in the industry; 2. Relevant education programs at both vocational and university levels providing both engineers and skilled labour capable of interacting *with* the engineers; 3. 'Infant industry'/'local content' policies; 4. Foreign oil companies funding Norwegian R&D and collaborating with Norwegian universities/institutes; 5. Technological challenges resulting in technological solutions that were adopted in most other offshore provinces; 6. The global acceptance of the relatively strict Norwegian safety regime benefiting Norwegian companies; 7. The market for ownership generally not being restricted; 8. Foreign actors buying Norwegian-owned companies and thus providing access to capital and global distribution channels; 9. Serial entrepreneurship, revenues from foreign takeovers reinvested; 10. Funding of petroleum-related R&D and promoting of collaboration between the government, academia, operators and suppliers.

In 2014/15, the bulk of Norwegian-based employment (around three-quarters) was related to the home market, and the large volume of employees led us to refer to this as 'Staff'. Development and provision of products and services for the export market ('Solutions') constituted slightly less than a quarter, while headquarters' functions of shipping and drilling companies ('Steel') constituted around 2 per cent of total employment.

We argued that the large volume of home market-based employment was due to: 1. A high volume of operations that require physical presence and where political/union pressure requires this presence to be carried out by Norwegian-based employees (drilling, well services, offshore maintenance, etc.); 2. Many cases where physical presence in operations is highly beneficial (after-market services, development and construction of customised offshore installations, R&D to solve challenges on the NCS, etc.).

We argued two important reasons for the large export activity: 1. Path dependency as Norwegian-based companies had developed solutions were taken in use globally and 2. Localisation in Norway providing benefits that would be hard to replicate elsewhere (collaborations in clusters, established, well-functioning project organisation, etc.). We argued that headquarters' functions for rigs and vessels were an important part of the industry even though such highly capital-intensive activity would never be very important for creating onshore employment.

So, what could the future hold for employment in the Norwegian-based petroleum supply industry, and what are the options for personnel with petroleum industry experience in activities outside the petroleum sector?

First, since most activities in the petroleum sector is based on the need for physical presence, and as there are clear benefits to physical proximity of operations, the eventual waning of Norwegian petroleum activity will lead to reduced employment in the home market-based supply industry. This could, in principle, lead to a shift in focus towards exports, but the possibilities for export related employment will be restricted by the following: 1. The issue of required physical presence; 2. Less ground-breaking field developments in the home market means fewer opportunities for developing new solutions at home; 3. Many of the successful export activities (for instance marine seismic and E&P software) are not as labour intensive as the construction of large platforms for the home market. The owner of a large supplier company states that it is simply not feasible to use Norwegian-employed personnel for the most labour-intensive activities in foreign markets:

> Our group is mainly based on services that tend to be performed 'on-site'. As Norwegians are generally not smarter than anybody else, we tend to hire locals when establishing offices abroad. Some of the group's companies do have technologies that could be supplied from Norway, but the resulting potential for employment is relatively small.
>
> (Owner of large supplier company)

Second, while novel use of digitalisation and automation (Baldwin, 2016) reduces the importance of physical presence and thus reduces the overall need for labour offshore, it could open opportunities for more export-related employment. The adoption of digitalisation and automation could to some extent be thwarted by union/political pressure to preserve national employment, however the potential for new export successes can be huge: first, novel digital solutions make it possible for more activities to be run or surveyed from a distance, e.g. from Norway. Second, Norwegian-based companies could use their edge in petroleum technology, including software development, to develop the needed digital solutions.

Third, the petroleum supply industry is to some extent a delicate ecosystem and if some central players, shipowners or certain large equipment suppliers for example, should move abroad due to gradually less activity on the Norwegian continental shelf, the benefits of the important clusters could be in danger.

In a future where neither proximity to a large home market nor 'local content' policies have much effect, employment in the Norwegian-based petroleum supply industry will gradually start to resemble the situation for employment in any 'normal' industry. In fact, it could start to resemble the employment structure in headquarters' functions for rigs and vessels, i.e. relatively small, technology-based units with both direct exports and export market activities run by foreign-based employees. There may thus be shift from 'Staff' to a combination of 'Solutions' and 'Steel'. In 2014/15, some large equipment suppliers were responsible for most of the exports from the Norwegian-based petroleum supply industry. One of our informants sees these companies as an historical anomaly, resulting from the sheer size of the NCS activity. The informant argues that the future will be dominated by small and medium-sized technology companies and headquarters.

> From today, being dominated by a few very large manufacturing companies, the future of export related employment is likely to be a large number of small and medium-sized technology companies and headquarters functions.
>
> (Former CEO of large well services company)

If the assumptions above hold, there would gradually be less Norwegian-based employment in the petroleum supply industry. What are then the options for more non-petroleum related employment? A foretaste of how this might evolve, has been observed in Norway in the period from the oil price collapse in the summer of 2014 until the petroleum sector downturn levelled out and started to rise again by the end of 2017. In the period 2014–2017, Norway saw personnel move or migrate to non-petroleum activities, either by oil sector employees going to work for non-petroleum employers in public or private sector, or petroleum supply industry employers (companies) exploring opportunities in other markets.

First, as the petroleum supply industry to a large extent employs personnel with generic competencies within engineering, finance, discipline technicians, etc., most of the employees are employable in other sectors, although some sectors may be reluctant to hire employees that have become used to the very generous two weeks on/four weeks off offshore rotation. The Norwegian unemployment rate rose after the crash in the oil price, but then gradually levelled off, proving that a large share of the former petroleum workers had been employed in other sectors.

Second, most of the petroleum supply companies are built on generic competencies, and in the aftermath of the oil price crash, a number of these companies tried partly or wholly to move into other markets. Shipyards that used to specialise in the construction (outfitting) of offshore vessels have turned to fishing, cruise and wind service vessels, offshore shipping companies provide service vessels for offshore wind farms, suppliers of electrical, instrumentation and automation solutions for the petroleum sector have won major installation contracts for large road tunnel projects, etc.

These are just a few examples of the versatile knowledge base that petroleum supply companies often possess. Be it opportunities in existing markets or from emerging demands, there is no rule that says personnel or supply companies from the petroleum sector should not succeed in addressing these opportunities. In addition, they will often bring with them petroleum sector experience and technology that can be applied or adapted to provide added value for their new customers. The foreign-owned companies with the most petroleum-specific technologies (say, drilling and well equipment) have so far been reluctant to venture into new markets through their Norwegian plants, but this is also said to be changing. However, as we have discussed in this chapter, the gradual build-up of the Norwegian petroleum supply industry based on existing skills and companies, took time and money, and so will large scale diversification into other markets.

Notes

1 See also: Been here since day one. Will be here tomorrow too. Schlumberger timeline poster for ONS [Offshore Northern Seas] 2002.
2 See: Key E&P technologies are invented and pioneered in Norway by GECO, WG and Schlumberger. Presentation by Schlumberger, 2015.
3 The Kongsberg Group is the successor to a state-owned weapons' factory. The state's 30 per cent stake in Aker Kværner was taken as a pre-emptive step when the industrialist Kjell Inge Røkke threatened to sell out to foreign owners (Ryggvik, 2013, 122).
4 It is not easy to define what constitutes a 'Norwegian-owned' company (Ryggvik, 2013) but at some point the merged Subsea 7 definitely ceased being 'Norwegian'.
5 www.hitecvision.com/about-us.
6 The standard systems of industrial classification (ISIC in the US and NACE in the EU) do not provide categories that allow for a precise demarcation of 'petroleum supply industry'. The NACE system includes a category called 'Support activities for petroleum and natural gas extraction' (NACE 9.1), but the building of platforms and rigs is found in NACE 30.113 and operation of offshore vessels in NACE 50.204. In addition, there are direct petroleum-related activities found in more general NACE codes. Marine seismic is included in NACE 71.122 'Geophysical surveys', and the major equipment suppliers (National Oilwell Varco, OneSubsea, MHWirth, etc.) are mostly included as NACE 28 'Manufacturing of machinery and equipment'. Between 1972 and 2003, the Norwegian labour market authority compiled annual surveys of employment in all sub-entities where upstream oil and gas activity was judged to be the main driver of activity. After the compilation of these surveys ceased in 2003, a number of consultancies/researchers have established their own mappings of Norwegian-based petroleum-related employment using similar definitions as those used by the Norwegian labour market authority (Rystad Energy, 2013a; Vatne, 2013, 2016).
7 www.norskoljeoggass.no/no/virksomheten/Arbeidsliv-og-kompetanse/Avtaler/Tariffavtaler/.

References

Aamodt, E., and Frøshaug, V. (2012). *Cluster as facilitator – The case of the maritime cluster in Møre and Romsdal.* (Master thesis), BI Norwegian Business School, Oslo.
Baldwin, R. (2016). *The Great Convergence. Information Technology and the New Globalization.* Cambridge/London: The Belknap Press of Harvard University Press.

Beyazay-Odemis, B. (2016). *The Nature of the Firm in the Oil Industry*. New York: Routledge.

Bjerke, H. K. (1989). *På dypt vann. Et høyteknologisk eventyr [In deep waters. A hi-tech adventure]*. Oslo: Ad notam forlag AS.

Bjørnstad, S. (2009). *Shipshaped. Kongsberg industry and innovations in deepwater technology, 1975–2007*. PhD thesis, BI Norwegian School of Management, Oslo.

Blomgren, A., Quale, C., Austnes-Underhaug, R., Harstad, A. M., Fjose, S., Wifstad, K., Melbye, C., Amble, A. B., Nyvold, C. E., Steffensen, T., Viggen, J. R., Iglebæk, F., Arnesen, T. and Hagen, S. E. (2015). *Industribyggerne 2015: En kartlegging av sysselsetting i norske petroleumsrelaterte virksomheter, med et særskilt fokus på leverandørbedriftenes eksportsysselsetning. [The builders of Industry 2015: A survey of employees in Norwegian petroleum related businesses, with a special emphasis on supplier industry employees related to exports]* Report IRIS 2015/031. Stavanger: IRIS.

Blomgren, A., Quale, C., Bayer, S. B., Nyvold, C. E., Steffensen, T., Tovmo, P., … Hagen, S. E. (2013). *Industribyggerne: Norsk olje- og gassnæring ut med havet og mellom bakkar og berg [The builders of industry: Norwegian oil and gas industry out by the sea and among hills and dales]*. Report IRIS 2013/031.

Bøe, A. E. (2012). *Slik at mennesker blir glade. Historien om Jon Gjedebo, skoletaperen som ble oljegründer og milliardær [To make people happy. The story about Jon Gjedebo, the school dropout who became oil entrepreneur and billionaire]*. Stavanger: Wigestrand forlag og Styrbjørn.

Engen, O. A. (2009). The development of the Norwegian petroleum innovation system: A historical overview. J. Fagerberg, D. Mowery and B. Verspagen (eds.), *Innovation, Path Dependency, and Policy: The Norwegian Case*. Oxford: Oxford University Press.

Gjerde, K. Ø. (2002). *'Stavanger er stedet' Oljeby 1972–2002 ['Stavanger is the place'. Oil city 1972–2002]*. Stavanger: Norsk oljemuseum.

Gjerde, K. Ø. (2012). Baseby for oljeletere [Supply base city for exploration companies]. G. Roalkvam and K. Ø. Gjerde (eds.), *Stavanger bys historie bind 4. Oljebyen 1965–2010*. Stavanger: Wigestrand forlag.

Grønås, T., and Hundsnes, L. (2002). Multilateral wells increse reserves, lower costs on Troll oil field. *Oil & Gas Journal* 00(23), 51–53.

Hanisch, T. and Nerheim, G. (1992). Norsk Oljehistorie. *Fra vantro til overmot? [The History of Norwegian Oil: From Disbelief to Arrogance?]* Oslo: Leseselskapet.

Hungnes, H., Kolsrud, D., Nitter-Hauge, J., Prestmo, J., and Strøm, B. (2013). *Ringvirkninger av petroleumsnæringen i norsk økonomi. Basert på endelige nasjonalregnskapstall for 2013 [The petroleum industry's ripple effects in the Norwegian economy. Based on final national accounts for 2013]*. Oslo: Statistics Norway. Report 2016/17.

Hunter, T. (2014). Law and Policy frameworks for local content in the development of petroleum resources: Norwegian and Australian perspectives on cross-sectoral linkages and economic diversification. *Mineral Economics*, 27, 11–126.

Isaksen, A. and Karlsen, J. (2012). What is regional in regional clusters? The case of the globally oriented oil and gas cluster in Agder, Norway. *Industry and Innovation*, 19(3), 249–263.

Krogh, F. E. (2002). 'Never mind!' Et møte med oljepionerene Johannes Viste og Svein Paulsen. ['Never mind!' A meeting with the oil pioneers Johannes Viste og Svein Paulsen.] *Norsk oljemuseum årbok 2002*. Stavanger: Norsk oljemuseum.

Lie, E. (2012). *Norsk økonomisk politikk etter 1905. [Norwegian economic policy after 1905.]* Oslo: Universitetsforlaget.

Nordbø, E. W. and Steinsland, N. (2015). *The petroleum sector and the Norwegian economy*. Oslo: Norges Bank Economic commentaries No. 4 2015.

Norwegian Academy of Technological Sciences. (2005). *Norwegian Petroleum Technology. A success story*. Trondheim: Norwegian Academy of Technological Sciences in co-operation with Offshore Media Group.

Pahr-Iversen, E. (2006). *1 pluss 1 kan bli mer enn 2 ... En fortelling om de første forsyningsbasene [1 plus 1 may be more than 2 ... A story of the first supply bases]*. Sola: Norsea Group.

Ryggvik, H. (2013). *Bulding a skilled national offshore industry*. Oslo: Norwegian Federation of Enterprise (NHO).

Rystad Energy. (2013a). *Aktiviteten i den petroleumsrettede leverandørindustrien i landets ulike regioner [Activity in the petroleum related supplier industry by region]*. Oslo: Rystad Energy.

Rystad Energy. (2013b). *Drivere og barrierer for teknologiutvikling på norsk sokkel [Drivers and barriers for technology development on the Norwegian shelf]*. Oslo: Rystad Energy.

Rystad Energy. (2017). *Internasjonal omsetning fra norske oljeserviceselskaper. [International sales from Norwegian oilfield services comppanies]*. Oslo: Rystad Energy.

Sasson, A., and Blomgren, A. (2011). *Knowledge Based Oil and Gas Industry*. Oslo: BI Norwegian School of Management Research Report 03/2011.

Tellnes, I., Ringbakken, K., Middelthon, T., Lilleheim, J. H., and Hamre, H. (2003). *Med havet som siktemål. Maritim utdanning i Stavanger gjennom 150 år. 1853–2003 [The ocean in sight. Maritime education in Stavanger for 150 years. 1853–2003]*. Stavanger: Stavanger Offshore Tekniske Skole.

Vatne, E. (2013). *Den spesialiserte leverandørindustrien til petroleumsvirksomhet. Omfang og geografisk utbredelse i Norge [The specialised supplier industry to the petroleum activity]*. Bergen: SNF Report 2013:2.

Vatne, E. (2016). *Sysselsetting i petroleumsvirksomhet 2015. Omfang og lokalisering av ansatte i oljeselskap og den spesialiserte leverandørindustrien [Employment in the petroleum sector 2015]*. Bergen: SNF Report 2016/03.

Additional sources

Economist. (08.10.2015). Norwegian Blues. *The Economist*.

Economist comments. (2015). www.economist.com/node/21672206/comments. Ref. the comment from guest-nmeosoa of 09:43, 12.10.2015.

EY (2018). *The Norwegian Oilfield Services Analysis 2017*: EY.

EY the Netherlands. (2015). *The Dutch oilfield services analysis 2014*: EY.

EY UK. (2016). *Review of the UK oilfield services industry*: EY.

Prosafe NOW. (2002). *Prosafe – a story about success. Interview with Reidar Lund. No. 1.* Retrieved 22.02.2018 www.prosafe.com/getfile.php/13259/PDF%20Filer/Prosafe%20NOW/Prosafe%20NOW%202002_1.pdf.

Part III

11 Versatile competences and product market diversification among oil and gas supply firms

Taran Thune and Tuukka Mäkitie

Introduction

Old reservoirs in the North Sea are now in a mature stage and recently discovered oil and gas (O&G) reserves are generally small or located under harsh conditions and hard to access localities. Moreover, climate change concerns create expectations for lower demand for fossil fuels in the long term. Issues like these have raised questions about the future of the O&G industry, and unsurprisingly, the question of diversification to new product markets has become an important issue in the Norwegian O&G industry and in industrial policy more generally. The high level of investment in advanced technological capabilities (both in terms of tools and skills) in the Norwegian O&G sector (see Chapters 2 and 3), are often mentioned as a key strength and source of future competitiveness for the industry. Moreover, the high skills, innovation capabilities and advanced technologies are expected to function as a stepping-stone for future opportunities and to enable firms to successfully apply their capabilities in new industries and product markets.

Another reason behind the belief in the transformative potential of the O&G industry is historical. This is not the first time that these firms have diversified into new markets. In fact, the Norwegian petroleum supply industry itself can be seen as an example of highly successful diversification of multiple companies and industries into the O&G market from the late 1960s onwards (see Chapters 2 and 10). Looking at their historical development (Ryggvik, 2010), it is easy to see that most large O&G supply firms have their roots in other industries such as exploitation of hydro power, shipbuilding and shipping (Engen, 2009), and more recently in ICT and advanced automation industries (Blomgren *et al.*, 2015). For instance, most of the large engineering, procurement and construction (EPC) companies (such as Aker and Aibel), companies that are involved in vessel design and construction (Vard, Ulstein) or in drilling (Archer, Seadrill) have a historical background in the Norwegian maritime industries dating back to the early twentieth century (Blomgren *et al.*, 2015; Engen, 2009; Ryggvik, 2010) (Chapter 10). Sub-suppliers of advanced equipment, such as equipment used in relation to drilling and on topside and subsea installations, often have a background in mechanical industry (such as National

Oilwell Varco, FMC Technologies, Scana, etc.). Moreover, many subsea companies have experience in electronics, sensor or communication technologies (Kongsberg Maritime, Roxar, ABB). Another example of a complex industrial history of petroleum suppliers comprises companies in the seismic industry (such as Petroleum Geo Services) and companies that provide advanced Exploration and Production software (Schlumberger) which draw upon industrial and academic competencies in geophysics, advanced ICT and maritime technologies.

These historical examples show that in the past, several firms in the petroleum supply industry have been able to transfer and adapt their prior capabilities developed in other industries to the needs of the growing O&G sector. This in turn has created expectations that also the O&G-related capabilities that the industry now possesses can in turn prove to be useful for future diversification. In other words, it is expected that O&G firms possess significant resources that can assist them in diversifying into new markets.

Despite this belief in the adaptability of the industry, and some case studies showing that O&G-related firms have diversified into marine sectors and to offshore wind power (e.g. Chapters 12 and 14; Mäkitie *et al.*, 2018; Steen and Hansen, 2014) there have been few systematic enquiries into the versatility of the petroleum competence base, or indeed the degree of diversification within the Norwegian supply industry. Limited knowledge also exists about the new markets to which suppliers have diversified, or which kind of competences and resources are more likely to enable diversification to other markets. Such questions are important, not only for a better understanding of the Norwegian petroleum supply industry, but also for policy makers seeking to diversify the industrial structure in countries with substantial investments in natural resources, and in particular hydrocarbon resources (Mahroum and Al-Saleh, 2017). The ambition of this chapter is to look into such questions, and through this contribute to the discussion regarding the future opportunities of O&G suppliers in new markets.

In this chapter, we draw upon the resource-based view of a firm (Barney, 1991; Penrose, 1959) as a theoretical perspective to approach the topic of industry diversification. According to Penrose (1959), the extent and types of resources (such as technological knowledge, human resources, managerial routines, physical infrastructure etc.) possessed by a firm influence if, how and where it diversifies. Firms are likely to seek economies of scale and scope by operating in similar or related markets where they can achieve synergies and growth through operations in multiple related markets (Montgomery and Hariharan, 1991; Rumelt, 1974; Teece, 1982). On the other hand, the transferability of resources between markets also plays an important role, as existing resources can be redeployed to new markets when seeking new growth opportunities (Sakhartov and Folta, 2014). While resources that are highly specialized for a single market (e.g. drilling in offshore conditions) are less likely to enable the application of these competences in other markets, more general competences (e.g. general engineering capabilities) are more likely to be transferable to other industries (Helfat and Lieberman, 2002; Levinthal and Wu, 2010; Pisano, 2016).

We apply these perspectives to investigate diversification patterns among Norwegian supply companies. We performed an empirical analysis of a sample of supply firms and their activities in petroleum and non-petroleum markets. In this analysis, we compared characteristics of diversified firms (operationalised as firms that offer products/services in other product markets in addition to O&G) to non-diversified firms in terms of the extent of resources (competence breadth and size of the company) and the types of competences (general-purpose and/or O&G specific) they possess. Moreover, we accounted for the relevance of age, economic performance and company profile for diversification in the supply industry. The data source was a database of 620 Norwegian supply firms where we applied the statistical method of logistic regression analysis and analysis of group differences.

We begin this chapter by reviewing the literature on firm diversification, emphasising the resource-based view as an explanation of such firm behaviour. Furthermore, we present the methodology of this chapter, followed by the empirical results. In the final section, we discuss our results in light of previous literature, and make concluding remarks regarding policy implications, limitations of this study, and further research.

Perspectives on firm diversification

There are several streams of research that have investigated characteristics of firms that diversify into new markets and the economic consequences of diversification. A core theory in this literature is the resource-based view of the firm, based on the seminal contributions of Edith Penrose. The resource-based view of the firm suggests that the competitiveness of firms is based on the possessed assets or resources of the company (Penrose, 1959; Wernerfelt, 1984). These resources include tangible resources such as technological artefacts, financial capital, facilities and physical infrastructure, and intangible resources such as different kinds of knowledge and managerial capabilities which are deeply embedded in the skill sets, technical systems and managerial systems (Barney, 1991). Such resources play a key role in how firms diversify to new product markets. Penrose's argument on the role of resources in diversification can be divided into two inter-linked perspectives.

In Penrose's first perspective, the extent or degree of available firm resources influences how likely the firms are to diversify. Firms always retain unused potential, or 'excess resources', which they can realise by diversifying into new markets. Firms can leverage e.g. their existing technological, design and customer knowledge in other markets, even though these resources have been acquired in other markets. Through economies of scale, firms can gain an advantage by, for example, sharing overheads and infrastructure over operations in several markets, thus being more competitive because of reduced production costs. Moreover, through economies of scope, firms are able to create synergies between activities in different markets which require similar resources (Nayyar and Kazanjian, 1993; Rumelt, 1974; Teece, 1982). In other words, due to similarities and complementarities between markets, firms are able to realize synergies between their operations by making use of existing expertise and resources, and develop a competitive advantage for themselves (Markides and Williamson, 1994). For instance, if different markets have similar customers, knowledge about these customers can be transferred within a firm. This creates a synergy in the firm from operating in several related markets, and can incentivize entry into a new market. Hence, diversification into related markets saves the firms the full extent of the cumbersome and risky acquisition of new knowledge (e.g. regarding customers) and assets needed in exploring unrelated markets (Rumelt, 1974).

In other words, due to opportunities of economies of scale and scope, a large abundance of resources is an important factor in diversification since such economies are more feasible when more resources are present. Firms with plenty of resources are in a position where they can diversify into more markets. Indeed, Montgomery and Hariharan (1991) found out that the breadth of the firm's resource base acted as a strong predictor of diversification. Moreover, Granstrand (1998) suggests that firms with high technological competences tend to diversify in multiple markets, and Wu (2013) finds that firms with larger stocks of innovation experience are more likely to diversify. For these reasons, we expect that abundance of resources is linked with diversification behaviour also in the case of the Norwegian supply industry. Stated differently, we expect that oil and gas supply firms with a lot of resources are more likely to be diversified than firms with fewer resources.

In Penrose's second perspective, also the types of possessed firm resources play a role in the decision to diversify (cf. Chapter 13). As mentioned above, firms are more likely to diversify if they are capable of applying their existing resources in another (related) market (Helfat and Lieberman, 2002). This is because, for instance, firms are more likely to identify opportunities in markets which have similarities to their existing markets. Benefits from related resources, for example in technology (Breschi *et al.*, 2003), products (Magnusson *et al.*, 2005; Tanriverdi and Venkatraman, 2005), markets (Nayyar and Kazanjian, 1993), manufacturing and assembly (Carroll *et al.*, 1996; Magnusson *et al.*, 2005), strategic management (Markides and Williamson, 1996; Prahalad and Bettis, 1986) and marketing and retail (Carroll *et al.*, 1996), have been identified.

However, not all resources are easily transferable to other markets. For example, previous technological experience in maritime operations is likely to help in identifying emerging business opportunities in offshore markets but less likely in on-shore markets. Key to the transfer of resources from one market to another is the specificity of resources (Pisano, 2016). This distinction means that resources have different degrees of transferability to different contexts. When resources are market-specific, firms usually focus on reaping value from exploiting the existing resources in markets in which they are already present (Teece, Pisano, and Shuen, 1997). In contrast, resources that are more generic or general-purpose enable better opportunities for diversification since these resources can be more easily redeployed and adapted to new situations (Levinthal and Wu, 2010). Therefore, the degree of market specificity of firm resources is likely to have a role in how much firms diversify. Stated differently, we expect that oil and gas supply firms with specialized oil and gas competences are less likely to diversify to other markets than firms with general-purpose (i.e. not oil-specific) competences.

In addition to the content of resources, their scalability plays a role in when firms diversify. By scalability we mean whether resources include an opportunity cost (such as applying a production facility to one purpose which disallows using it simultaneously for other purposes), or not (e.g. using codified knowledge, like patents, which can be applied in several purposes at the same time) (Levinthal and Wu, 2010; Wu, 2013). In this perspective, market fluctuations such as growth and decline in demand create incentives for firms to redeploy their resources to their most profitable use between markets. However, the timing issue of diversification remains outside the scope of this chapter, and is further discussed in Chapter 13 of this book.

Methodology

The basis for the empirical analysis of the supply industry is a database we have constructed that contains information on Norwegian petroleum supply firms. The selection criteria of the companies in the sample (an estimated total of about 2,500 petroleum service and supply companies exist in Norway) was that they were members of one of several industry associations for petroleum-related companies in Norway, and through this define themselves as petroleum suppliers. By using membership registries, we were able to collect a sample of 620 companies from a range of public sources, such as public accounts data, company databases and information available on the company websites or in news media.

To shed light on the question of how diversified the supply industry is, we have collected information on the different markets where petroleum supply firms offer products or services. This information has been collected through company websites, product brochures and similar marketing materials. To code a company's presence in different 'product markets' we collected explicit statements about the presence of a company in different product markets. We are aware that information might in some cases only signify an ambition to move towards this market, and is not a measure of the volume of activities in alternative markets.

In cases where the companies had their headquarters abroad, we looked only at the activities of the Norwegian branch. We collected information about all markets where products or services were deployed in a 'bottom-up' fashion, i.e. without predefined categories. Later, similar categories were merged to make a more feasible analysis. We have used the data to measure the range of alternative markets in which the companies are active (defined as 'market breadth'). In the analysis below, we use this as a dependent dummy variable, comparing the diversifiers and the non-diversifiers, i.e. companies that only offer products/services in the O&G market vs. companies that also offer services/products in at least one other market.

To investigate which characteristics of firms are most likely to influence the probability of product market diversification, we look at several independent variables (Table 11.1). In terms of our first expectation about a positive relationship between resource extent in firms and degree of diversification, our first independent variable measures the breadth of the competences in a firm. This data was collected by looking at company information in the national registry of Norwegian companies about NACE codes,[1] NACE subject descriptions and the publicly announced purpose of the company. Moreover, to triangulate this data and to add competences not included in NACE codes, we gathered information from all available company websites of our sample. We used a predefined classification scheme for key competences in the supply industry for coding purposes (Blomgren, 2016). We constructed the competence breadth variable by summing the number of different competence areas a firm possesses, and dividing the firms into four categories. We also looked at the size of the company, measuring the number of employees in each company (three categories in total) as a second indicator of the breadth of available resources in the firm. We assume that large firms have more resources, not only financial, but also in terms of competences.

The independent variable to address our second expectation was the market specificity of the competences which measures whether a firm had either general-purpose competences (related, for example, to ICT, construction, engineering, mechanical and electrical competences that can be applied across multiple sectors) or competencies that are highly specific to O&G operations (i.e. related to drilling and well operations, and geology and reservoirs). To classify these competences into the two categories, we utilized a classification scheme for the O&G industry (Blomgren, 2016) regarding competence bases in the petroleum-related industries.

We also looked at several other variables in the analysis. First, we used a simple typology of firms that classifies firms into two categories: product (i.e. firm offering an artefact or otherwise concrete products) and service companies. Other variables included are company age and the average annual economic revenue of the companies over the last ten years (or the number of years since registration if less than ten).

To investigate the relationships between our variables, we performed a logistic regression where product market diversification is used as a binary dependent variable and the other variables are used as predictors (Table 11.1). The analysis performed assessed the probability of being a diversified company (having products in at least one other market than oil and gas) for firms with different characteristics.

Table 11.1 Variables used in the analysis

Variable	*Values*	*Frequency*	*Per cent*	*Data source*
Product market diversification	0 = only products/services in O&G market 1 = products in at least one alternative market	0 = 271 1 = 345	44 56	Company websites, product sheets
Competence breadth	1–4 or more competence areas	1 = 80 2 = 330 3 = 109 4 or more = 99	13 54 17 16	NACE-codes, purpose statements, company websites
Size of company	The number of employees (average or one specific year) 1 = less than 50 2 = 50–250 3 = 251 and above	1 = 280 2 = 152 3 = 76	45 30 15	Public registries
Market specificity of competences	0 = general purpose competences 1 = O&G specific competences	Broader definition of O&G competences 0 = 372 1 = 243 Narrow definition of O&G competences 0 = 466 1 = 149	60 40 76 24	Classification scheme
Company profile	1 = Product company 2 = Service company	1 = 201 2 = 412	33 67	Purpose statements, company websites
Age of company	1 = before 1990 2 = 1990–2000 3 = after 2001	1 = 153 2 = 150 3 = 270	27 26 47	Public registries
Economic performance	Average annual profits last 10 years, or average by number of years of income if less than 10 years old 1 = less than NOK 1 M per annum (pa) 2 = NOK 1–3 M pa 3 = NOK 4–6 M pa 4 = NOK 7–10 M pa 5 = above NOK 11 M pa	1 = 308 2 = 132 3 = 47 4 = 36 5 = 93	50 21 8 6 15	Public registries

Results of the empirical analysis

In our sample, 44 per cent of supply firms were not present in other markets than O&G, but 56 per cent were. From these diversified firms, 40 per cent were diversified into one or two other markets besides O&G, and every fifth company was diversified into three or more product markets.

As can be seen in Figure 11.1, petroleum suppliers have activities in a range of other markets in addition to O&G. The most common was the maritime industry, including shipping and transport sectors. Moreover, land-based manufacturing industry (including metal and process industries) was another common alternative market. In addition to these two main alternative markets, 63 firms were involved in the construction sector while more than 30 firms also provided products and/or services related to energy supply, clean technologies, defence and aviation as well as environment and safety.

A total of 60 per cent of firms could be categorized as having a single field of competence, while 27 per cent had two or three in total. Only 13 per cent of firms had four or more types of competence fields. However, there was a wide variation in competences. Unsurprisingly, products and services related to drilling and oil- and gas-well operations, as well as competences related to the offshore and marine industry, were most common. Also, a number of companies were specialized in supply and transport functions of offshore installations and platforms.

Several suppliers possessed more general-purpose competence, for instance in mechanical fields such as machinery, piping and metal works (see Figure 11.2). Reflecting the high-tech side of modern offshore industries, more than a fifth of

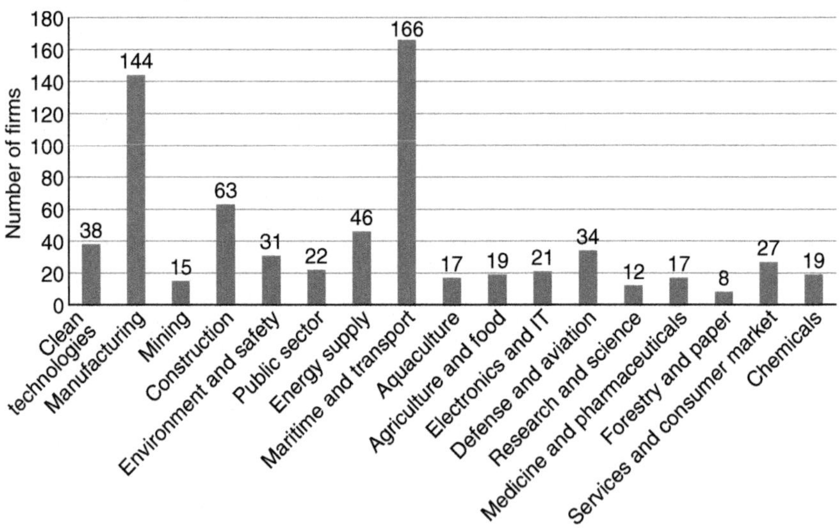

Figure 11.1 Non-O&G markets of supplier firms ($n = 616$).

Source: own database.

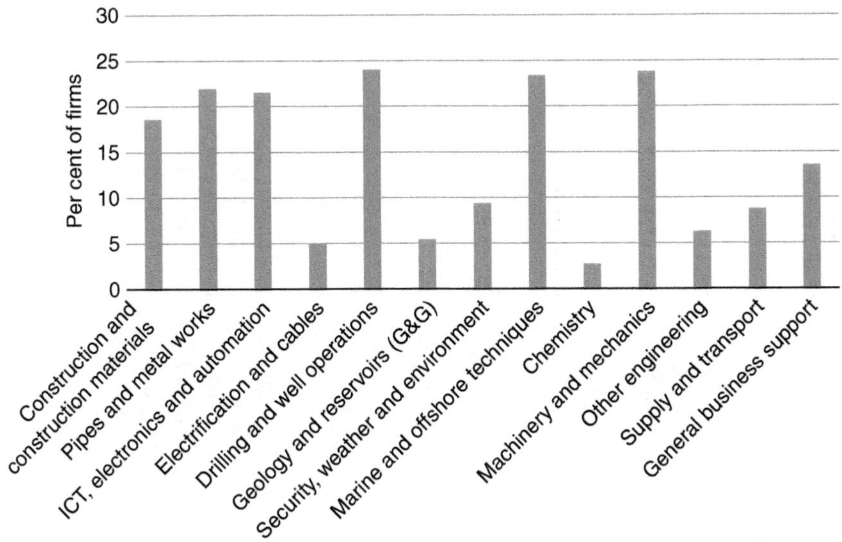

Figure 11.2 Competence bases, percentages of firms having the competence ($n = 616$). Classification based on Blomgren *et al.* (2017).

the firms had competences in ICT, electronics and automation. Moreover, almost 20 per cent of firms were in construction or building material businesses. The O&G supply industry has also attracted firms with, for instance, security, and environmental protection and electrification competence, and general businesses support functions like human resources services and business consulting (categorized under 'general business support').

Having product market diversification as the dependent variable, Table 11.2 summarizes the results of the logistic regression analysis. The regression model of the independent or predictor variables on product market diversification was statistically significant overall, and the model correctly classified 70 per cent of the cases (compared to 57 per cent without the independent variables), but slightly more than 80 per cent of the cases of diversification. The model estimates the probability of being a diversified company for companies with specific sets of characteristics. For each variable, the probability is compared to a reference category. Both the coefficients and the odds ratio for each category are described.

As can be seen in Table 11.2, competence breadth had a negative relationship with diversification. This entails that supply companies having competences in several knowledge fields are not associated with being present in other markets than O&G. In other words, supply companies with a large variety of competences are not more diversified than companies with a narrower range of competences. This is contrary to our expectations derived from theory.

Also in terms of our other indicator measuring the relationship between resource breadth and diversification, large companies (measured in terms of the

Table 11.2 Logistic regression on dependent variable product market diversification (presence in non-O&G markets)

Variable names	*Coefficients (ExpB in parenthesis)*	*Odds ratio (ExpB) to be diversified in comparison to reference category*
Competence breadth: 1 area as reference category		
Competence breadth: 2 areas of competence	–1.800* (0.165)	80% lower odds (to be diversified)
Competence breadth: 3 areas of competence	–1.044* (0.352)	60% lower odds
Competence breadth: 4 or more areas of competence	–0.425* (0.654)	35% lower odds
Size of company (50 or less employees is reference category)		
Medium (51–250 employees)	–0.277* (0.758)	24% lower odds
Large (more than 250 employees)	–0.237 (0.789)	Lower odds (but not significant)
Market specific vs. general competences (General competences as reference category)		
Maritime and/or O&G specific competences (broad definition of O&G competences)	–0.427* (0.653)	35% lower odds
Only O&G specific competences (narrow definition of O&G competences)	–1.891* (0.151)	85% lower odds
Company profile (service firms as reference category)		
Type of company – product company	0.546* (1.727)	73% higher odds
Age (Established before 1990 as reference category)		
1991–2000	0.380* (1.463)	46% higher odds
After 2000	0.136 (1.146)	Higher odds (but not significant)
Average annual profits (less than 1 mill. is reference category)		
1–3 mill.	0.401* (1.494)	50% higher odds
4–6 mill.	–0.084 (0.920)	Lower odds (but not significant)
7–10 mill.	0.971* (2.639)	164% higher odds
More than 11 mill.	–0.462* (0.630)	37% lower odds
Constant	1.587 (4.891)	
Pseudo R^2 (Nagelkerke)	0.277	
Observations	616	

Note
* significant at 0.5 level.

number of employees) are not significantly related to diversification. Again, contrary to our expectations, it had signs of negative relationship since medium-sized firms were less likely to be diversified than small firms. However, as the odds ratio for this relationship was small, and the category 'large companies' was insignificant, we must conclude that size had an ambiguous relationship to diversification in this sample.

We also used two different measures to test the differences of possessing general competences versus O&G specific competences in relation to the probability of being a diversifier. The first and broader definition of O&G specific competences included 1) drilling and well operations, 2) geology and reservoirs as well as 3) marine and offshore techniques, while the second narrower definition included only the competence areas 1 and 2. Both definitions of O&G specific competencies show a significant negative relationship to diversification with the most O&G specific competences (i.e. the narrow definition) being least likely to be diversified to other markets. In other words, companies with highly specific O&G-related knowledge were less likely to be diversified.

In terms of the other variables in the regression, product-oriented firms are significantly more diversified compared to service firms. This result suggests that the profile of the firm has relevance for diversification. The age of the firm, on the other hand, was not a significant predictor of diversification overall, but younger firms were more likely to be diversified. The variable 'economic performance' showed a mixed picture. Two groups of companies were more likely to be in the group of diversifiers – companies with low average revenues, and companies with comparatively moderate revenue (but not the top performers) were both more likely to be diversified compared to the smallest companies in terms of average revenues. This analysis indicates that young, small and not very affluent companies with generic competences have a higher probability of being in the group of supply firms that have diversified.

Discussion and conclusions

This chapter has looked into diversification behaviour of petroleum supply companies in Norway. The petroleum industry has a history of attracting diverse industrial competences into O&G, and several of the large supply and service companies today emerged in other industries and predate the Norwegian oil history by decades – in some instances by centuries. This historical diversification pattern into oil has been seen as providing a good basis for further diversification out of oil and gas when petroleum resources eventually run out or the demand for oil is significantly reduced. With this in mind, this chapter has presented an empirical study of diversification of petroleum supply companies which has shed light on the type of market the firms have diversified into, the competences they possess and what characterises diversified and non-diversified petroleum supply companies.

We drew upon the resource-based view of a firm to guide the analysis of firm diversification behaviour. The resource-based view of firm diversification postulates that firms which have an extensive resource base and general-purpose resources (in comparison to market-specialized resources) are more likely to diversify (Montgomery and Hariharan, 1991; Penrose, 1959; Pisano, 2016; Teece, 1982). We used these insights to formulate two expectations regarding the characteristics of diversified companies, and subjected these to an empirical test using logistic regression analysis.

The analysis indicated that the majority of petroleum supply firms in the sample had some activities in other markets than O&G. Our results suggest, however, that O&G supply firms tend to diversify into a relatively limited number of other markets, i.e. to the related markets where they presumably are able to apply their existing resources, such as technological competences and previous business experience, with ease. Shipping and transport (including the maritime sector), land-based manufacturing (such as process industries) and the construction sector were the three most common additional markets. The concentration of diversification efforts, especially in the first two markets, is remarkable in the light of the relatively broad base of competences existing in the supply industry (presented in Figure 11.2). This finding would suggest that despite the breadth of competences in the O&G supply industry, these firms have not diversified to a high number of alternative markets.

This finding is reflected in our statistical analysis which does not show support for the expectation that having a large resource base encourages diversification. Quite to the contrary, we find that breadth in the competence base of supply companies was linked to less product market diversification. At the same time, we find that company size, measured by the number of employees, showed an insignificant relation to the probability of being diversified. These results lead us to reject the expectation that supply companies with more resources would be more diversified than supply firms with fewer resources. One reason that likely explains this is that large supply companies function as 'systems integrators' (Acha, 2002). Their core competencies concern designing complex solutions that integrate a range of technologies, rather than developing specific lines of products that can be marketed in several markets. This means that even though they maintain a broad portfolio of technological capabilities, their unique skill in integrating different solutions is highly market specific.

We suspect that this finding can also be related to the unique munificence of the O&G market. This market offers highly profitable business opportunities for supply firms, especially during market booms. This can encourage firms to focus on the O&G market since it provides high relative demand and opportunities in comparison to other markets (cf. Wu, 2013), and firms therefore diversify by entering new niches within petroleum and building competencies in multiple specialities, rather than moving to new markets outside oil and gas. Moreover, we note that many of the large companies had mainly market-specific assets that they had deployed in the still lucrative O&G market. In other words, we see that many suppliers were indeed diversified, but the large and technologically advanced companies did not seem to be leaders in this process. An implication of this finding, assuming that diversification is seen as a desirable societal outcome, is that deliberate policy interventions should focus on supporting the relatively small and not very resourceful companies in their diversification efforts.

Our results further show the importance of types of firm resources for diversification behaviour. In our sample, firms with competences which are highly specific to the O&G market, related for example to drilling and managing oil reservoirs, were least likely to have activities in other markets. Hence, we are able to confirm the second expectation in suggesting that highly O&G specialized companies are less likely to be diversified into other markets.

This finding has implications for the study and policy of diversification of the O&G supply industry. First, the notion of differentiating between general-purpose and market-specific resources helps to identify competences and firm resources which can be more easily redeployed into other markets (Pisano, 2016). Hence, in firm-level strategy-making and public policy addressing diversification, future competence building should prioritise the development of more general-purpose competences (see Chapter 12). This is pertinent because such competences can 'create options for future market entry and they can be complementary to market-specific capabilities' (Pisano, 2016, 23). Second, this notion of general-purpose and market-specific resources enables further study and policy-making to identify companies that have the highest opportunities to diversify, and therefore enables more efficient industrial transformation policy targeting value chain segments with general-purpose competences.

To conclude, we reflect on the limitations and opportunities of further study. First, as shown by significant statistical relationships between product market diversification and company type (product or service firms), age and economic performance of firms, the above-mentioned extent and types of resources are not the only important factors describing characteristics of diversified firms. Second, due to our snap-shot type of analysis of firm characteristics, we are unable to address the time-specific co-occurrence of variables. As diversification remains a significant policy challenge in countries with significant petroleum resources (or other natural resources, cf. Mahroum and Al-Saleh, 2017), we call for further research addressing the above perspectives in studies of the potential for diversification of natural resource industries.

Note

1 In innovation studies, NACE-codes are often used as an indicator of firms' knowledge bases. While the O&G industry is not represented as an individual NACE-code, we found the standard classification code system also gave very limited insight on the subject matter.

References

Acha, V. L. (2002). *Framing the Past and Future: The Development and Deployment of Technological Capabilities by the Oil Majors in the Upstream Petroleum Industry*. PhD thesis, Brighton, UK: University of Sussex.

Barney, J. (1991). Firm resources and sustained competitive advantage. *Journal of Management*, 17(1), 99.

Blomgren, A. (2016). *Høyere teknisk utdannelse med relevans for petroleumsvirksomheten* Arbeidsnotat IRIS 2016/399. Stavanger: IRIS.

Blomgren, A., Quale, C., Austnes-Underhaug, R., Harstad, A. M., Fjose, S., Wifstad, K., Melbye, C., Amble, A. B., Nyvold, C. E., Steffensen, T., Viggen, J. R., Iglebæk, F., Arnesen, T. and Hagen, S. E. (2015). *Industribyggerne 2015: En kartlegging av sysselsetting i norske petroleumsrelaterte virksomheter, med et særskilt fokus på leverandørbedriftenes eksportsysselsetning. [The builders of Industry 2015: A survey of employees in Norwegian petroleum related businesses, with a special emphasis on supplier industry employees related to exports]* Report IRIS 2015/031. Stavanger: IRIS.

Breschi, S., Lissoni, F., and Malerba, F. (2003). Knowledge-relatedness in firm technological diversification. *Research Policy*, 32(1), 69–87.

Carroll, G. R., Bigelow, L. S., Seidel, M.-D. L., and Tsai, L. B. (1996). The fates of de novo and de alio producers in the American automobile industry 1885–1981. *Strategic Management Journal*, 17(Summer), 117–137.

Engen, O. A. (2009). The development of the Norwegian petroleum innovation system: a historical overview. J. Fagerberg, D. C. Mowery, and B. Verspagen (eds.), *Innovation, Path Dependency and Policy. The Norwegian Case*. Oxford: Oxford University Press.

Granstrand, O. (1998). Towards a theory of the technology-based firm. *Research Policy*, 27(5), 465–489.

Helfat, C. E., and Lieberman, M. B. (2002). The birth of capabilities: market entry and the importance of pre-history. *Industrial and Corporate Change*, 11(4), 725–760.

Levinthal, D. A., and Wu, B. (2010). Opportunity costs and non-scale free capabilities: profit maximization, corporate scope, and profit margins. *Strategic Management Journal*, 31(7), 780–801.

Magnusson, T., Tell, F., and Watson, J. (2005). From CoPS to mass production? Capabilities and innovation in power generation equipment manufacturing. *Industrial and Corporate Change*, 14(1), 1–26.

Mahroum, S., and Al-Saleh, Y. (eds.). (2017). *Economic Diversification Policies in Natural Resource Rich Economies*. Abingdon: Routledge.

Mäkitie, T., Andersen, A. D., Hanson, J., Normann, H. E., and Thune, T. M. (2018). Established sectors expediting clean technology industries? The Norwegian oil and gas sector's influence on offshore wind power. *Journal of Cleaner Production*, 177, 813–823.

Markides, C. C., and Williamson, P. J. (1994). Related Diversification, Core Competencies and Corporate Performance. *Strategic Management Journal*, 15, 149–165.

Markides, C. C., and Williamson, P. J. (1996). Corporate Diversification and Organizational Structure: A Resource-Based View. *The Academy of Management Journal*, 39(2), 340–367.

Montgomery, C. A. and Hariharan, S. (1991). Diversified expansion by large established firms. *Journal of Economic Behavior & Organization*, 15(1), 71–89.

Nayyar, P. and Kazanjian, R. (1993). Organizing to attain potential benefits from information asymmetries and economies of scope in related diversified firms. *The Academy of Management Review*, 18(4), 735.

Penrose, E. (1959). *The theory of the growth of the firm*. Oxford: Blackwell.

Pisano, G. (2016). *Towards a Prescriptive Theory of Dynamic Capabilities: Connecting Strategic Choice, Learning, and Competition*. Harvard Business School Working Paper, 16–146.

Prahalad, C. K., and Bettis, R. A. (1986). The dominant logic: A new linkage between diversity and performance. *Strategic Management Journal*, 7(6), 485–501.

Rumelt, R. P. (1974). *Strategy, structure, and economic performance*. Boston, Mass: Harvard University Press.

Ryggvik, H. (2010). *The Norwegian Oil experience: A toolbox for managing resources?* www.sv.uio.no/tik/forskning/publikasjoner/tik-rapportserie/Ryggvik.pdf. Retrieved 22.02.2018.

Sakhartov, A. V., and Folta, T. B. (2014). Resource relatedness, redeployability, and firm value. *Strategic Management Journal*, 35(12), 1781–1797.

Steen, M., and Hansen, G. H. (2014). Same Sea, Different Ponds: Cross-Sectorial Knowledge Spillovers in the North Sea. *European Planning Studies*, 22(10), 2030–2049.

Tanriverdi, H., and Venkatraman, N. (2005). Knowledge relatedness and the performance of multibusiness firms. *Strategic Management Journal*, 26(2), 97–119.

Teece, D. J. (1982). Towards an economic theory of the multiproduct firm. *Journal of Economic Behavior & Organization*, 3(1), 39–63.

Teece, D. J., Pisano, G., and Shuen, A. (1997). Dynamic capabilities and strategic management. *Strategic Management Journal*, 18(7), 509–533.

Wernerfelt, B. (1984). A Resource-Based View of the Firm. *Strategic Management Journal*, 5(2), 171.

Wu, B. (2013). Opportunity costs, industry dynamics, and corporate diversification: Evidence from the cardiovascular medical device industry, 1976–2004. *Strategic Management Journal*, 34(11), 1265–1287.

12 Diversification into new markets

Challenges and opportunities for petroleum supply firms

Allan Dahl Andersen and Magnus Gulbrandsen

Introduction

This chapter analyses diversification processes in Norwegian firms that have the petroleum industry as a major market but have initiated efforts to serve customers in other industries. As seen in Chapter 11, many petroleum supply firms have attempted diversification to other markets. This chapter looks further into what characterises processes of diversification and it is based on interviews primarily with firms – some of them with successful diversification processes, others with attempts with unclear results.

In recent years the petroleum industry has been exposed to at least two forms of transformation pressure that motivated supply firms to enter other markets, i.e. to diversify. First, in recent years the sector has suffered under relatively low oil prices, particularly affecting high-cost regions like the North Sea. This has compelled firms to cut costs in their main market (as described in Chapter 5) but also to find new sources of revenue (as discussed in Chapter 11). Second, there is an emerging pressure from climate change mitigation and a longer-term perception that petroleum is an industry in decline. Although this may not necessarily imply an imminent economic risk for the supply firms, they experience enhanced uncertainty concerning long-term prospects of this market. In Norway, there is strong attention to both of these pressures, and 'moving beyond oil' has been central in the public debate for a lot longer than the most recent crisis (e.g. seen as the most important issue by the voters in the 2013 parliamentary election, cf. Aftenposten, 2013).

Our main goal is to understand the challenges that firms encounter when they attempt to diversify away from petroleum. In our perspective, diversification is an innovation process that involves entering a new industrial market. From the innovation literature, it has repeatedly been emphasised that such processes are challenging because they necessitate breaking with well-established organisational routines and practices (Nelson and Winter, 1982). We expected firms primarily to approach diversification incrementally by modifying existing products to fit the perceived needs in new markets. For example, we assumed that a producer of valves for petroleum installations would search for opportunities in other markets where valves are used and, if necessary, implement changes to the valves, their production and marketing. We also expected that the firms would start with markets where the needs were perceived to be most similar to oil and gas.

These expectations were only partly met. Although many of the firms in our sample had indeed approached diversification in such an incremental and step-wise way, most of them emphasised that the challenges were much more fundamental than what could be solved by incremental 'tinkering'. Interviewees often emphasised stark contrasts between the industries in which they tried to gain a foothold and the petroleum context, particularly with respect to institutional and systemic conditions (as described in Chapter 2), not so much the technologies. For some firms, this implied long-lasting transformations of many aspects of their organisations, including marketing, production, and research and development (R&D). But there are many nuances to this picture, contingent on aspects such as the size of firms, their position in the value chain, their technology/products and the timing of diversification.

In the next section, we briefly review some important features of the practical and theoretical sides to diversification of petroleum suppliers. We see diversification as a process whereby a firm tries to match its current capabilities to new and partly unknown external conditions. Understanding this process is theoretically interesting and of high policy relevance. The following section presents the main methodological considerations with emphasis on interview data coding and analysis. The next section contains the main empirical analysis. Finally, in the last section we discuss contributions and practical implications. Two important points are the disruptive nature of organisational rather than technological changes, and the perceived insufficiency of traditional innovation policy instruments to support the form of organisational diversification with which these firms struggle.

Theoretical aspects of diversification

The theoretical and practical question underlying our interest in the process and challenges of diversification in supply firms is whether and how the vast resources – technologies, capabilities, human capital and so on embedded in petroleum-oriented products and services – can be redeployed in other economic activities to stimulate broader industrial diversification (e.g. Andersen *et al.*, 2015; Mäkitie *et al.*, 2018). In addition, we are interested in how such activities can be supported by policy.

Diversification by firms refers to entering new markets and introducing novelty. Theoretically, we view diversification as an evolutionary innovation process that is new in the market dimension and possibly also in the organisational and technology dimension (Abernathy and Clark, 1985). As with other innovation processes, diversification is systemic and open (involving different actors), path-dependent (set in specific trajectories) and influenced by technological, social and political developments inside and outside the firms (Chesbrough, 2003; Dahlander and Gann, 2010; Helfat and Lieberman, 2002).

In this chapter, we limit our understanding of diversification to the process through which established firms enter new markets rather than new product segments within existing markets (or industries, we use these terms interchangeably in the context of diversification). We follow Helfat and Lieberman (2002) in understanding patterns of diversification primarily as a matter of matching a firm's resources and capabilities accumulated in the 'old' industry to the requirements of the new. In this perspective, such inter-industry differences define the challenges of and barriers to diversification.

Different concepts have been suggested for understanding this matching process such as 'capability stretching' (Wang and Chen, 2015) and 'creative accumulation' (Breschi *et al.*, 2000). These concepts indicate the degree to which existing resources and capabilities can be redeployed in the new industry and at what cost (Farjoun, 1994; Helfat and Lieberman, 2002). A common distinction is between related and unrelated diversification (Rumelt, 1974). Unrelated diversification is more challenging, and therefore rarer since firms need to acquire a substantial portion of new resources and capabilities and integrate these with existing ones (Helfat and Peteraf, 2003; Kogut and Zander, 1992). The associated effort required by diversifying firms is thought to be high and equivalent to disruptive innovation. Related diversification on the other hand is seen as involving a more modest need for integrating new resources and capabilities and therefore only incremental innovation efforts by firms. It is therefore the most commonly observed pattern of diversification, although it is natural to perceive relatedness as a scale rather than as a binary construct.

Relatedness between industries is also multi-dimensional, although most research emphasises the importance of *technological* relatedness. Besides technology (Breschi *et al.*, 2003), it can be linked to market properties (Tanriverdi and Venkatraman, 2005), capabilities for production (Magnusson *et al.*, 2005) or capabilities for innovation management (Helfat and Lieberman, 2002). The matching process must therefore take place in several, and often interdependent, dimensions which may mean that resources and capabilities are related in some dimensions but unrelated in others.

Last, we also know that diversification often is a long-term process which despite product innovation and market entry may not yield commercial benefit for long periods (Cantwell and Vertova, 2004). This indicates that costs and benefits are distributed inter-temporally, which can imply significant financial strain. This point also highlights the importance of the length of time needed to acquire and integrate new capabilities in a firm. Information about how firms view the relationship between petroleum and the new/target industries and about internal and external conditions for diversification is therefore needed. We expect that incremental innovation/related diversification does not pose major complications. However, in situations of low relatedness, firms may experience larger challenges. Finally, diversification is linked to the broader issue of industry transformation, which means that the role of policy instruments in firms' diversification processes is of interest.

Methods

In this chapter we will explore the characteristics and challenges of diversification processes by use of a qualitative method. This enables us to tease out some of the underlying dimensions and variations in diversification processes and the contexts in which they occur. We have broadly opted for a grounded theory approach (Glaser and Strauss, 2017), rather than developing a strict theoretical framework. We have used conceptual considerations from the literature to design interview questions, but at the same time allowed new analytical dimensions to emerge from the data with open-ended questions that made room for probing and exploring respondents' own terms and understandings.

We operationalised our conceptual considerations in two stages. First, we asked the firms about their main activities in petroleum in order to acquire an indication of the firms' core resources and capabilities. Second, we asked a set of nuanced questions regarding what the firm did or needed to do differently so as to enter the new industry, and what were experienced as the most difficult challenges. These questions highlighted resource and capability gaps.

We assessed degrees of relatedness between industries as the discrepancy between the resources and capabilities firms had accumulated in petroleum and those needed to succeed in the new industry. In turn, the extent of such discrepancy across dimensions of relatedness is indicated by what firms considered most challenging in the diversification process. For example, if firms did not mention technology when asked about the main challenges, we interpreted this as stating that technological relatedness is relatively high and not a major problem.

In our coding of interviews, we developed an integrative synthesis between our main concepts and the data, resulting in five analytical categories. The theoretical concepts worked well as structuring devices and we categorised the data accordingly. These conceptual categories include mode of innovation management including R&D and new product development, organisation of production, market properties (including contracting, financing and other contextual aspects), temporal aspects and, lastly, policy issues. For each category, we considered differences between firms based on characteristics such as size, dependency on oil and gas, types of technology and position in the value chain.

Our main data material consists of 18 interviews of which 14 were with petroleum supply firms, two with client companies and two with industry experts. The intention with the latter four interviews was to gain insight into the broader conditions for diversification and to assess how clients in new markets view diversifying firms in petroleum. We selected firms that historically have had petroleum as their main market and have recently attempted to establish themselves in other industries in order to gain insight into practical experience with diversification. We identified these 'lead diversifiers' through news items in the Norwegian press, further snowballing – asking the first interviewees about other firms with similar experiences. In addition, we selected firms that have diversified into different types of industries – for example, offshore wind power, aquaculture and medical technologies – as well as firms that represent different segments of the petroleum supply chain. In this manner we strived to find firms and interviewees that represent a variety of experiences and contexts. Interviews were semi-structured, and carried out between December 2016 and April 2017, with an average duration of 50 minutes. The most frequent types of informants were Chief Executive Officers (CEOs) and Chief Technology Officers (CTOs). All interviews were taped, transcribed and loaded into the NVivo qualitative research software. Table 12.1 provides an overview of interviews and various details of the firms. Each interview is ascribed a numeric code which we use in the text when referring to interviews as sources of information.

Findings: main challenges for diversifiers

In this section we analyse how firms in our sample express experiences with diversification. All have attempted – some with a fair degree of success – to diversify. Most of them have done this by modifying their existing products (tubes, safety suits, steel structures etc.) for new markets, or by approaching new markets with a more generalised perspective on their services – '*Our core competence is managing huge projects, really, not just in petroleum*' (I-3) as stated by an interviewee of an engineering firm. But all firms had experienced differences between petroleum and their new market(s) – differences which often challenged fundamental aspects of the organisation and its business model. Table 12.2 summarises the main findings, which then are described more thoroughly in the next sections.

Innovation management: product and technology development

Petroleum is – or perhaps was – characterised by an advanced but predominantly responsive mode of technology development. Petroleum firms approach suppliers with specific problems that they want solved under a specific set of circumstances. Suppliers establish innovation projects focused on satisfying these pre-expressed demands. This process is typically not only initiated but also financed by the petroleum firms. For example, for new oilfield development, the supply firms and the client go through a four-stage process where only the last step implies cost estimates that can form the basis of a clear contract. The earlier stages oriented towards selecting and improving overall design (e.g. platform or drilling ship) are financed by the petroleum companies, and several firms in our sample expressed that they have significant leeway in these design stages.

The described organisation of technology development was not found in any of the new markets that the firms aimed at. In other markets such as offshore wind power and aquaculture, suppliers are expected to develop a product internally that complies with relevant standards, and only then approach potential buyers. In other words, a higher 'technology readiness level' is required in these industries. Interviewees stated that many of the activities of technology development, including prototype testing and certification, have to be paid for internally and carried out prior to intensive interaction with clients. Interviewees generally found the new industries more risk-averse and less open to radically

Table 12.1 Overview of interviews and firm information

Interview no.	*Value chain position/interviewee type*	*Core product/competence*	*Year for diversification*	*Phase of diversification*[1]	*New market(s)*
I-1	Procurement, construction and installation	Steel structures offshore, system integration	2003	3	Offshore wind
I-2	Subsea equipment and installation	Cable protection systems	2010	4	Offshore wind
I-3	Procurement, construction and installation	Platforms, steel structures	2008	3	Offshore wind
I-4	Subsea equipment and installation	EPC, system integration	2015	1	Aquaculture, offshore energy
I-5	Procurement, construction and installation	Ship design and building	2005	4	Offshore wind
I-6	Support, operational and other services	The production of textile safety equipment	2010	3	Offshore wind
I-7	Subsea	Waste management solutions	2012	2	Organic waste management
I-8	Support, operational and other services	Product design consultancy	2014	2	Medicine
I-9	Subsea equipment and installation	System integration	2015	3	Defence & aquaculture
I-10	Engineering	ICT control systems/ cybernetics	2014	3	Solar power, smart housing, tunnels
I-11	Topside and processing equipment	System integration and hoses and hose fittings for all purposes.	2013	4	Aquaculture
I-12	Topside and processing equipment	Filter solutions	2014	4	Aquaculture
I-13	Subsea	Expertise on cables used in harsh environments.	2013	2	Aquaculture, defence, renewable energy
I-14	Topside equipment	High-voltage, small, and durable electronics.	2013	4	Electrical vehicles, space industry
I-15	Renewable energy project operator	n.a.	n.a.	n.a.	n.a.
I-16	Renewable energy project operator	n.a.	n.a.	n.a.	n.a.
I-17	Subsea industry association	n.a.	n.a.	n.a.	n.a.
I-18	Energy consultancy firm	n.a.	n.a.	n.a.	n.a.

Note

1 We define phases as follows. Phase (1): firm is scoping options for diversification but is yet to act on them. Phase (2): firm has commenced R&D search projects that could help penetrate new market. Phase (3): firm has made initial sales but still infrequent and small-scale contracts. Phase (4): firm is relatively established in new market with steady and significant activity.

Table 12.2 Main differences between petroleum and new markets

Relatedness dimension	*Petroleum*	*New markets*
Innovation management	User-driven innovation R&D financed by client Tech exploration within projects Customized design R&D goal: customization and durability	Strategic innovation R&D at own expense Tech exploration prior to projects Standardized design R&D goal: low unit cost
Organization of production	Small-batch production Customized production Maximize quality Design production line for high quality Engineering & production integrated	Large-batch production Serial manufacturing Minimize unit price Design production line for high volume Engineering & production sequential
Market properties	Customer confidentiality Few & similar customers Few, large, and long contracts Conventional finance High profit margins	Arm's length relation with customers Many & diverse customers Many, small(er), and short(er) contracts Project finance Low(er) profit margins

new concepts. Apart from the clear economic and organisational challenges, some interviewees also claimed that this influenced innovativeness.

One example is a shipbuilding firm which expected that potential customers within offshore wind power would be interested in offering a contract based on a conceptual ship design and the good reputation of the firm, but this was not the case. The customers offered no guarantees and support, and the firm would need to build the ship first and then return to the customer. Banks were not interested in funding a 'weird boat' either, because the lifetime value could not be estimated in a traditional way, and because there was no guarantee from the customer. The informant elaborated (I-5):

> You will not get new concepts going until you have done all testing yourself. In oil and gas, at least before the 2014 crisis, many had such large profit margins that they were willing to try out new things. It was much easier to introduce new things in oil and gas than in offshore wind.

Overcoming this type of challenge implies a shift from what we may term a user-driven form of technology development to a more strategic, austere and internally driven form of development. Almost all interviewees described this as a huge problem with setbacks, trial-and-error with limited impact, and in some cases reductions in R&D personnel. The renewable energy firms stated that supply firms which contacted them were often 'poorly prepared' in terms of products and ideas about value added (I-16), confirming the challenges that these firms have when approaching new markets.

Another and related difference in technology development relates to the notion of customisation. Search heuristics guiding technology development in petroleum often address problem-solving particular to the natural environment as is often the case in natural resource-based industries (Andersen and Wicken, 2016; Andersen *et al.*, 2018). Each reservoir provides a particular set of challenges related to, for example, going deeper and further and with unique safety requirements. The petroleum sector has institutionalised a strong emphasis on safety because failure can mean oil spills, fatalities or ecosystem damage. This further drives customisation, use of high quality materials, high cost-levels, and places extensive demands on documentation. As expressed by a technology manager of an engineering firm (I-3), '*Everything is one of a kind [in petroleum]. We never make copies. This is because each oil reservoir is totally unique and therefore you never need the same [equipment].*' The high degree of customisation implies that technology development is an integrated part of each project rather than an activity between projects and contracts.

Again, the contrast with most other industries was clear, where the logic in most cases was cost reduction through standardisation of technology components to facilitate economies of scale. In offshore wind power, for example, less emphasis was put on quality and safety, and much more on price. The interviewees stated that less severe consequences of equipment failure and a lower degree of maturity (less developed regulation) were central explanations.

Some interviewees nevertheless pointed to similarities between petroleum and other industries. One firm highlighted the similar emphasis on customisation and documentation requirements within medical technologies and defence contracts due to safety concerns; another that some products for the petroleum sector are standardised as well, '*though new designs for [workers'] suits are required by regulators and customers every 3–5 years, a suit is still a suit, and these are mass-produced in a standardised production line*' (CTO, textile firm) (I-6). A shipbuilding firm stated that the demands for customisation were much higher in some settings such as offshore wind than in the more mature petroleum sector (I-5). Nevertheless, for most firms the standardisation and price dimensions further challenge the internal organisation of innovation. The search heuristics of R&D efforts must be changed more towards efficient production of standardised items – something which is a new exercise for most of the firms in our sample.

Organisation of production

Related to the high degree of customisation, contracts in the petroleum industry most often involve small batches rather than a large volume. This has important implications for how firms organise the production process. To accommodate customisation in petroleum, firms must have a large degree of flexibility in production. Flexibility comes from many manual and engineering hours that raise costs. R&D is, in other words, typically integrated with production making it possible to experiment with product design so that a high-quality output can be achieved. In the new markets, however, focus was on minimising unit prices through mass-production of standardised components.

Firms articulated this as a change from engineering towards serial manufacturing. In the words of a manager of an engineering firm (I-1):

> [A] major difference is that offshore wind is about serial production rather than small-batch production. For offshore wind, you often deliver 15–50 identical items while for petroleum it is normally 1–3 items per project. For offshore wind you therefore need more focus on planning the production steps and [to] ensure that component stocks and logistics are in place.

This change also often includes attempts at a more efficient organisation of production. The CEO of a subsea company (I-13) explained that to compete in the new market they had to '*build a totally new production line, and worked a lot with 'LEAN' thinking in our overall organisation and production to bring down cost.*'

An important aspect of such reorganisation is to separate engineering and production more clearly. For example, a subsea firm (I-2) stated that it used to be '*50% engineering and 50% production*', but after moving into offshore wind it had become '*more like 20% engineering and 80% production*'. The firm said that '*this reduced labour input by 25% per output unit*'.

Large-batch production also has repercussions for innovation management. One manager of a subsea company explained:

> If you can change your component design such that production per unit is 10 minutes faster, then you can make money because these 10 minutes per unit are valuable when you produce thousands of units. In petroleum, engineering was more about tailoring solutions to each project.
>
> (I-2)

Many firms experienced the changes required in the organisation of production as very challenging and most indicated that this adaption is a long-term and ongoing process.

Market properties

From long-term coordinated to shorter-term multi-party contracts

The typical form of contracting between supply firms and their clients differs between petroleum and most other industries. We highlight four aspects: type of contracting, size and duration of contracts, culture of billing, and funding of project development.

In petroleum, so-called EPC (engineering, procurement and construction) contracts dominate. These imply that the client has one to two interfaces with turnkey supply firms – systems integrators – which manage a network of sub-suppliers. Many of the firms are therefore part of these coordinated networks and have close contractual relationships with only a few actors. By contrast, 'multi-contracting' is the most common form of contracting within offshore wind power. Here, the client manages many interfaces – up to 20 different partners – directly, which interviewees claim is because it allows them to have more control of the projects so as to reduce costs. For the supply firms this is unattractive because risks are higher and channels of communication more complex. Moreover, there are many more potential clients, and competition for their attention is fiercer.

Interviewees stated that the most attractive petroleum contracts last five years, much longer than for example wind park installation which typically is six to nine months. A newcomer with a limited portfolio of orders can have costly, in-between-project periods of inactivity. In addition, diversifying firms need to operate in a more heterogeneous market with more frequent sales activities. This means that more organisational resources go into customer care, marketing and sales.

Furthermore, the culture of billing is different. The typical pattern described in petroleum is that cost overruns (concerning agreed price) are normally not problematic and firms are allowed to send an extra invoice. In other sectors, the price is fixed from the beginning of a contract and additional costs become the supplier's own problem. Moreover, timing and speed were described as more flexible in petroleum. If a supplier delivers a good quality product within 'reasonable time', a generous price can be asked. In sectors such as construction and offshore wind power, delays are penalised with day fines, requiring the firms to be much more disciplined.

Finally, in petroleum there is funding available for comprehensive 'pre-study' analyses and informal and interactive learning between suppliers and their clients. In other industries, funding is different. For example, offshore wind most often involves 'project financing' in which creditors demand strong influence on project planning and commissioning in the form of a meticulous and low-risk plan, and a framework for aspects such as risk allocation and time discipline that were seen as 'rigid'. Interviewees claimed that this led to 'very conservative' technology choices with only 'proven technology' being implemented rather than more innovative solutions. This complicates diversification in two ways for firms in our sample. First, it makes it difficult to acquire the first contract without a proven track record in the new market; second, firms have less support from customers in the process.

Customer relations

The informants made many additional statements about how petroleum differs from their new markets. Most seemed to agree that petroleum was more 'mature' than in many other markets. By this, they referred to different characteristics such as the level of technology, the standardisation of procurement and competition for contracts, and the strict national/international safety standards and strong public involvement in regulation. A few described this maturity in a more negative way, claiming that it also implied contractual and market relationships that were close to monopolistic. A CEO from an ICT firm, for example, stated that the petroleum industry '*is not interested in new actors. It is extremely difficult to get inside due to rigidity in the industry*' (I-10). Others stated that '*low maturity cold be beneficial*'. One CEO, for example, remarked: '*Aquaculture is like a teenage industry: a bit Wild West with few rules but many opportunities*' (I-12).

As mentioned, the new markets were generally regarded as more competitive and with different types of business-to-business marketing and sales activities than responses to procurements. Most of the firms seemed to believe that a 'less open' mode of innovation was required when they diversified into other industries. Petroleum was characterised by intense and long-term relationships based not just on contracts but also on trust with relatively few actors, while in other industries relationships were more arms-length based on sales of internally developed products and services. This type of industry difference suggests the need for a rather different approach to customer relationships which seems to be difficult to adopt. Indeed, several firms struggled to get into good working relationships with clients in other industries and to understand needs in the new market. Some stated that they were still in an information-gathering phase making new contacts through participation in trade fairs and similar events. For several firms, diversification starts with broader involvement at public and open gatherings.

Temporal aspects of diversification

A general finding, especially among firms that only started to diversify in recent years, is that it is easy to underestimate the time needed for diversification. A technology manager from an electro-mechanical products firm (I-13) emphasised that the biggest challenge '*is that things take time, and time is money*'. Although the process of learning about new clients and their needs, and adapting and marketing products and services to these needs, was not necessarily difficult in a technical or social sense, he stated that this took much time, and thus personnel resources. Others indicated that redefining the firm as something else than a petroleum supplier had led to tensions between different parts of the firm (the board and top management). One stated that the main barrier for diversification was '*internal strategic decisions*' and that the firm's leadership was too 'conservative' (subsea company; I-4). Another informant stated that most supply firms '*have not been willing to embrace diversification, partly because they were waiting for better times in petroleum*' (I-10).

Moreover, most firms in our sample are still in the process of finding their feet in the new market (only five of 12 firms are in phase 4 of diversification, see Table 1) while experimenting with organisational changes including innovation management, production lines and customer relations. Hence, it is not yet certain that maintaining their activity in the new area will prove financially viable in the longer term.

Because diversification requires significant resources and takes time, the issue of timing becomes important. In the view of one informant: '*If you decide to diversify only when you are in a crisis, it will be too late*' (I-11). The same informant observed that many petroleum suppliers have the ability and willingness to diversify but have been too busy with surviving the petroleum crisis to dedicate the necessary attention and resources. Struggling simultaneously in new and old markets is problematic for a more evolutionary process of organisational learning.

The relatedness between industries seems to change over time. Some informants stated that conditions for succeeding in petroleum are changing as well: '*When there was money in petroleum we got funding for development: this never happens in aquaculture. (...) But it doesn't happen in oil and gas anymore either*' (technology manager, mechanical products firm, I-11). Indeed, several firms saw possible synergies between the recent need to reduce costs in petroleum and the likelihood of succeeding in non-petroleum markets. Some informants also highlighted that low maturity of new markets is an advantage. One CEO argued that '*if you produce milk, you can't just start producing meat and expect to outcompete established slaughterhouses. There are more opportunities in emerging, niche markets*' (I-11). Hence, although often characterised by high uncertainty, diversification to involve emerging product markets may be easier due to less competition from established players.

Policy issues

The majority of firms seemed to be at an early stage in the diversification process and expressed diverging views on public support. The difference between getting development costs paid in full by a petroleum client and receiving 18 per cent tax deduction for internal development costs was perceived as dramatic by one company, calling for policymakers to 'do something'.

Some had received funding from the Research Council of Norway and several had projects under the R&D tax deduction scheme SkatteFUNN (details on schemes in Chapter 2). However, several firms emphasised that the adequacy of these policy instruments depends on context. For example, if there is a need to diversify due to a crisis in the main market, traditional instruments work too slowly. In the words of one R&D manager: '*It is OK but takes too long and requires too much energy. You haven't even gotten a response on your application before you are passed the concept design phase (in a project)*' (I-7). On the other hand, if the diversification strategy is well-planned, such instruments are helpful. Several also mentioned use of funding from the Work and Welfare Authorities for retraining of staff.

Nevertheless, the general message from firms was that they felt that Norwegian policymakers had done too little to support diversification. It is unclear whether this reflects the policy portfolio of public support for innovation and industrial development or (as can be seen in some interviews) a reflection of a 'crisis management mode' with little time to learn about new venues of collaboration and support. It can be claimed that what the firms seem to miss is dedicated support for diversification rather than more general support for various types of innovation, R&D and (re)training.

Conclusions and implications

Petroleum supply firms that have tried to enter other industries have faced a number of challenges. The main ones stem from a low level of relatedness between petroleum and other sectors. As expected from theory and empirical research on relatedness between industries, this implies challenging processes of matching existing competences, structures and processes to a new setting. For most firms, technology does not seem to be the main issue – perhaps because they seem to choose target markets for diversification based on perceived relatedness in the product or technology dimensions. However, the 'softer' aspects related to organisation of innovation, engineering and production, types of contracts and funding of R&D, clearly constitute more fundamental or disruptive challenges for the firm.

The firms are accustomed to bespoke technology development and production processes, often oriented towards large-scale one-of-a-kind outcomes, generously supported by the petroleum clients and carried out in a system with few but intensive linkages. For most firms, the new industries operate differently: they have to streamline production, increase in-house R&D and technology development, and maintain linkages to a much larger number of potential customers. This transformation is very challenging, and is made harder by the petroleum downturn and the emphasis on cutting costs.

It should be added that this is a somewhat stylised view of petroleum and the target industries of the firms. Some firms mass-produce for petroleum firms; others experience unique and complicated markets elsewhere such as in healthcare and defence (as described in Chapter 11). A related issue, touched upon in Chapters 2 and 3 of this book, is the degree to which petroleum represents a unique industry, or perhaps whether the special characteristics mentioned by many of the supply firms are also strongly related to the cyclical nature of the industry and the earlier strong upturn with high oil prices (as described in Chapter 5).

We have conceptualised diversification as an innovation process, i.e. a process that is long-term and prone to dynamic and often constraining processes such as lock-in and path-dependency. The validity of such a conceptualisation has been confirmed by the interviews. Moreover, it is conceptually interesting that our case portrays how different dimensions of relatedness interact. This is a topic which merits more attention in the future.

Many of the firms have used various public sources of support for R&D, innovation and retraining of personnel, but they still express that they would have wished for more policy mechanisms dealing directly with diversification. As technology is not the main challenge, we would expect R&D policy instruments to be inappropriate and inadequate for supporting diversification processes. This relates both to the timing and content of diversification where, perhaps, experimental policy instruments supporting development of new organisational capabilities might prove useful. Finally, some firms highlighted that it is relatively easy to succeed in a new market if it is immature and emerging due to the lack of established players. Support for diversification could thus equally address the demand side of innovation, for example by supporting the formation of new technological or market niches with high growth potential.

We have analysed the experiences of a few 'lead diversifiers', and our results suggest that the topic merits more attention. Later investigations may proceed in different ways, for example by studying diversification processes longitudinally or by making a broader comparative study through targeting a much larger number of firms.

References

Abernathy, W.J. and Clark, K.B. (1985). Innovation – Mapping the Winds of Creative Destruction. *Research Policy*, 14, 3–22.

Andersen, A.D., Johnson, B., Marín, A., Kaplan, D., Lundvall, B.-Å., Strubrin, L. and Kaplinsky, R. (2015). Natural resources, Innovation and Development. *Globelics Thematic Review*, Aalborg, Aalborg University Press.

Andersen, A.D., Marín, A. and Simensen, E.O. (2018). Innovation in natural resource-based industries: a pathway to development? *Innovation and Development*, 8.

Andersen, A.D. and Wicken, O. (2016). Natural resource knowledge idiosyncrasy, innovation, industry dynamics, and sustainability. Working Papers on Innovation Studies. Centre for Technology, Innovation and Culture, University of Oslo.

Breschi, S., Lissoni, F. and Malerba, F. (2003). Knowledge-relatedness in firm technological diversification. *Research Policy*, 32, 69–87.

Breschi, S., Malerba, F. and Orsenigo, L. (2000). Technological Regimes and Schumpeterian Patterns of Innovation. *The Economic Journal*, 110, 388–410.

Cantwell, J. and Vertova, G. (2004). Historical evolution of technological diversification. *Research Policy*, 33, 511–529.

Chesbrough, H. (2003). *Open Innovation*. Boston, MA, Harvard Business School Press.

Dahlander, L. and Gann, D.M. (2010). How open is innovation? *Research Policy*, 39, 699–709.

Farjoun, M. (1994). Beyond Industry Boundaries: Human Expertise, Diversification and Resource-Related Industry Groups. *Organization Science*, 5, 185–199.

Glaser, B.G. and Strauss, A.L. (2017). *The discovery of grounded theory: strategies for qualitative research*, London: Routledge.

Helfat, C.E. and Lieberman, M.B. (2002). The birth of capabilities: market entry and the importance of pre-history. *Industrial and Corporate Change*, 11, 725–760.

Helfat, C.E. and Peteraf, M.A. (2003). The dynamic resource-based view: capability lifecycles. *Strategic Management Journal*, 24, 997–1010.

Kogut, B. and Zander, U. (1992). Knowledge of the Firm, Combinative Capabilities, and the Replication of Technology. *Organization Science*, 3, 383–397.

Magnusson, T., Tell, F. and Watson, J. (2005). From CoPS to mass production? Capabilities and innovation in power generation equipment manufacturing. *Industrial and Corporate Change*, 14, 1–26.

Mäiktie, T., Andersen, A.D., Hanson, J., Normann, H.E. and Thune, T.M. (2018). Established sectors expediting clean technology industries? The Norwegian oil and gas sector's influence on offshore wind power. *Journal of Cleaner Production*, 177, 813–823.

Nelson, R. and Winter, S.G. (1982). *An evolutionary theory of economic change*, Cambridge, MA: Harvard University Press.

Rumelt, R.P. (1974). *Strategy, structure, and economic performance*. Boston, Mass: Harvard University Press.

Tanriverdi, H. and Venkatraman, N. (2005). Knowledge relatedness and the performance of multibusiness firms. *Strategic Management Journal*, 26, 97–119.

Wang, T. and Chen, Y. (2015). Capability Stretching in Product Innovation. *Journal of Management*, 44(2), 784–810.

Additional sources

Aftensposten. (21.08.2013). Norge etter oljen, klimaendringer og asylpolitikk er viktigste saker [Norway after the oil, climate change and refugee policy are important topics]. Aftenposten (News media).

13 From oil to wind, and back again

Resource redeployment and diversification

Tuukka Mäkitie, Taran Thune and Jakoba Sraml Gonzalez

Introduction

Climate change concerns, controversial and technologically challenging extraction areas, and price competitiveness of renewable energy sources all forecast an eventual decline of the oil and gas (O&G) industry in Norway. As this industry is an important provider of jobs and welfare in the country, new industries are needed to compensate this likely decline of economic activity in the forthcoming decades. As described in previous chapters, a potential diversification process is not without costs and challenges, which is why most firms attempt to diversify towards what are perceived as related markets. Diversification to related markets (i.e. markets where competences and capabilities of the industry can be applied) therefore present an opportunity for petroleum economies to develop new industries by using prior industrial experiences.

Several O&G firms in Norway have diversified to technologically-related offshore wind power (OWP) industry (Hansen and Steen, 2015). However, this diversification has proven to be far from a straightforward process (see Chapters 11 and 12). While similarities between markets in, for example, offshore technologies and installation techniques have been utilized in entering the OWP market, dissimilarities in demand conditions, market opportunities and institutional set-up have hindered the full commitment of many diversified O&G firms in OWP (Chapter 12; Mäkitie *et al.*, 2018). Moreover, the engagement of the O&G industry in OWP has been intermittent, with more activities in OWP during declines in the O&G market, and less during an O&G market boom period (Mäkitie *et al.*, 2017).

While fluctuations in commodity prices are a common phenomenon in natural resource industries, their role in facilitating diversification to related industries has been less discussed. Literature on diversification often theorizes that economies of scope and scale (i.e. synergies and complementarities from having activities in multiple markets) explain why firms engage in diversification (e.g. Montgomery and Hariharan, 1991; Tanriverdi and Venkatraman, 2005), but temporal differences in engagement in diversification, and the role of changes in market demand, have been less elaborated.

This chapter aims to advance the knowledge about the phenomenon of intermittent diversification by looking more closely at the role of firm resources, such as technological knowledge and managerial capabilities. More specifically, we analyze how the *scalability* of redeployed firm resources can be connected to intermittency in diversification behaviour in changing market conditions. Moreover, we argue that *investments* in market-specific resources are likely to be linked to more sustained engagement in diversification.

To address these issues, we investigated the diversification behaviours of five O&G 'lead diversifiers' in the OWP market: Statoil, Kværner, Aibel, Siem Offshore and Ulstein. We analyzed the diversification of these firms during 2007–2016 in the light of market developments in O&G and OWP. The analysis is based on secondary data sources such as news media articles, annual reports, press releases and financial data. We analyzed the characteristics of resources which were redeployed across markets, and the prevailing conditions in the marketplace at the time of resource redeployment (Levinthal and Wu, 2010; Lieberman *et al.*, 2017; Sakhartov and Folta, 2014). We approached this task with a two-step process. First, we took note of the scalability of redeployed firm resources in diversification as well as investments in market-specific resources. Second, we analyzed how the firms' engagement in diversification had changed over time in different market situations.

We commence by briefly reviewing literature regarding resource deployment between markets. We follow this discussion by presenting the methodology of this study and by introducing our case study. We then present our analysis of the diversification patterns in the five companies. In the final section, we discuss the implications for industrial policies intended to support the growth of new industries through diversification.

Scalability of resources and investments in related markets

We understand firms as consisting of varied resources such as technological competences, managerial capabilities, human resources, facilities and other types of capital goods which firms apply to create value (Penrose, 1959). Since resources are difficult to imitate, they are regarded as the basis for firm competitiveness (Barney, 1991; Teece, 1982).

When firms diversify, they are more likely to enter related markets where they can apply their existing resources and can accomplish synergies (Rumelt, 1974; Tanriverdi and Venkatraman, 2005). Market relatedness allows firms to *redeploy* resources from one market into a new market with limited costs (Sakhartov and Folta, 2014). Because of this, firms can apply some of their existing resources and therefore will have fewer sunk costs in the new market. This means that the firms have less risk in entering a new market (Sakhartov and Folta, 2015). In turn, having lower risks reduces barriers for diversifying, and enables entering and exiting a related market relatively quickly. Hence, diversification in related markets with possibilities to redeploy existing resources creates an opportunity to experiment in a new market with an option of 'easy' retreat (Lieberman *et al.*, 2017).

The patterns of entry and exit behaviour in related markets can depend upon the types of resources that the firm redeploys. *Scale-free resources*, such as knowledge, patents, brands and customer relationships, can be used simultaneously for different purposes, and can be shared between different markets without significant opportunity costs (i.e. losing an alternative to use the same resource elsewhere at the same time). In comparison, *non-scale-free resources* such as human and financial resources, machinery and facilities have limits in their applicability as they cannot be used simultaneously for multiple purposes and thus can cause an opportunity cost (Levinthal and Wu, 2010; Wu, 2013).

Since they incur different opportunity costs, scale-free and non-scale-free resources are likely to differ in how they are redeployed in different market situations. Since underutilized capacity becomes available when opportunities in a market do not meet with the extent of available resources (Levinthal and Wu, 2010), firms can seek to redeploy non-scale-free resources from markets with less demand into more profitable market(s). Hence, the *relative demand* and differences between markets of a diversified firm play a key role in redeployment of non-scale-free resources (Helfat and Eisenhardt, 2004). For deployment of scale-free resources, such relative demand between related markets plays a lesser role because of fewer opportunity costs. Therefore, the *absolute demand*, or business opportunities in a market which are independent of comparable opportunities in other markets, is more important for a firm deciding to enter or exit a related market with its scale-free resources (Wu, 2013). In other words, scale free resources are less likely to be redeployed back-and-forth between markets according to relative changes in markets. The opposite is true for non-scale free resources as their redeployment can more easily lead to intermittent engagement in diversification during periods of volatile market conditions.

While possibilities to redeploy resources between markets can create benefits for the company, differences between markets nevertheless always exist. For instance, even technologically related markets often differ in terms of business logics, and therefore the management of a diversifying firm needs to learn or otherwise acquire management capabilities that are suitable for successfully operating in the new market (Prahalad and Bettis, 1986). In other words, to be competitive in the diversified market, it is often necessary for firms to invest in market-specific resources (Pisano, 2016). They generally have two options for this: to invest in specialized resources internally (e.g. in-house R&D, retraining personnel), or externally (e.g. firm acquisitions) (Lee and Lieberman, 2010). Especially large companies with substantial financial and managerial resources can choose to develop new resources through internal expansion where they may utilize their existing resources (Helfat and Lieberman, 2002). However, external acquisition can provide a faster way to acquire the needed resources.

Either way, such investments are likely to solidify the engagement in diversification even if market conditions change. This is because investments in specialized resources create sunk costs in the new market, as such resources might be lost in cases where the diversification does not go as expected. This means that firms become more committed to the new market and are not as likely to abandon a diversified market, even if the relative demand in markets changes and the core market (e.g. O&G) becomes more attractive again (Dixit 1989; Lieberman *et al.*, 2017).

To sum up, while a diversification strategy focusing on redeploying resources in related markets comes with a lower risk and reduced barriers for entry and exit (Lieberman *et al.*, 2017), disregarding investments can also be an unsuccessful diversification strategy in the long run. Merely exploiting existing resources is unlikely to lead to long-term benefits as it does not improve the competitive position of the firm over time in an evolving market (Markides and Williamson, 1994). Hence, a long-term and consistent diversification effort would include both redeployment and investment in new resources (Pisano, 2016). In Table 13.1 we sum up the above notions in terms of factors that are linked to patterns of consistent or intermittent engagement in diversification in changing market conditions.

Methodology

We performed a longitudinal case study analysis of the diversification activities of five companies in the Norwegian O&G industry. We selected these firms because they represent important supply-chain segments in the O&G industry and are early movers in diversification to OWP. The firms are all large and established enterprises with O&G as their main market. Statoil stands out as the only petroleum operator and by far the largest company in the sample. The study covers the period from 2007 to 2016.

The focus of our analysis is on the diversification of the firms in OWP and how the engagement in OWP varied over time. In terms of the latter, we looked at whether the OWP activities involved redeployment of scale-free or non-scale-free resources. We also took note of any identifiable investments in new OWP-specific resources. Moreover, we analyzed how the entry and possible exit patterns from OWP coincided with market and demand situations in O&G and OWP.

We used several data sources to perform the analysis. We searched archives of news data regarding public statements, investments and contracts made by these companies in the OWP market. Using the Retriever/Atekst tool, we performed a news data search in five major media sources reporting on events in the Norwegian O&G industry. Moreover, we analyzed annual reports and press

Table 13.1 Factors linked to intermittency of engagement in diversification during changing market conditions

Consistent engagement	*Intermittent engagement*
• Redeployment of scale-free resources (no opportunity costs involved) • Investments in specialized resources (sunk costs in the diversified market)	• Redeployment of non-scale free resources (opportunity costs involved) • No investments in specialized resources (no sunk costs in the diversified market)

releases of the companies regarding diversification in OWP, and the overall economic development of the firms in the study period by using public accounts data. We framed the firm-level analysis with an industry-level presentation of O&G and OWP market developments based on secondary literature and data from Statistics Norway (SSB) and Wind Europe (see Mäkitie *et al.*, 2017 for more details on methods and data).

Market developments and the engagement of oil and gas industry in offshore wind power

The oil price influences investments in oil extraction by shaping expectations for the profitability of developments. Due to the technologically demanding natural conditions, maturity as an oil-bearing shelf and move into frontier areas, low expectations in oil price often translate into fewer investments in the Norwegian Continental Shelf, and vice versa. Indeed, since 2007, both oil price and investment levels have been in flux. Brent oil price had increased to more than US$100 per barrel until the financial crisis in 2008 causing a crash in price and a dip in investment levels during the period 2008–2010. However, by 2011 the oil price had already recovered to a high-level, leading to a boom in investments until 2014. The oil price then collapsed again leading to yet another downturn period during 2015–2016 with a dramatic fall of investments from a record level of NOK 220 billion in 2014 to NOK 162 billion in 2016. In sum, the level of absolute demand in the O&G market has been fluctuating throughout 2007–2016, with fewer business opportunities during downturns and many opportunities during the boom period.

Due to e.g. high availability of hydropower and discontinuous political support for OWP farms, there has not been a domestic market for OWP in Norway (Normann, 2015). Therefore, Norwegian OWP firms, large numbers having originated in the O&G industry, are directed towards international markets, especially in Europe (Normann and Hanson, 2017). Since 2005, the European market has grown rapidly, fuelled especially by the development of large wind farms in the UK and Germany (Kern *et al.*, 2014; Wieczorek *et al.*, 2015). Driven by policy measures to increase the production of renewable energy in Europe, the OWP market has developed into a significant line of business. In fact, in 2016 the European OWP market surpassed that of the O&G market in the Norwegian Continental Shelf with the total investment of €18.2 billion (NOK 169 billion). In sum, demand in the OWP market has grown steadily during 2007–2016, making it an increasingly attractive market for diversifiers with redeployable resources. Consequently, during O&G downturn periods, the OWP market has had an attractive relative demand, while O&G boom periods with very profitable opportunities have reversed the situation in favour of the O&G market.

In an earlier study we showed that the overall engagement of the Norwegian O&G industry in the OWP market has fluctuated over time in line with changes in relative demand (Mäkitie *et al.*, 2017). As can be seen in Figure 13.1, the

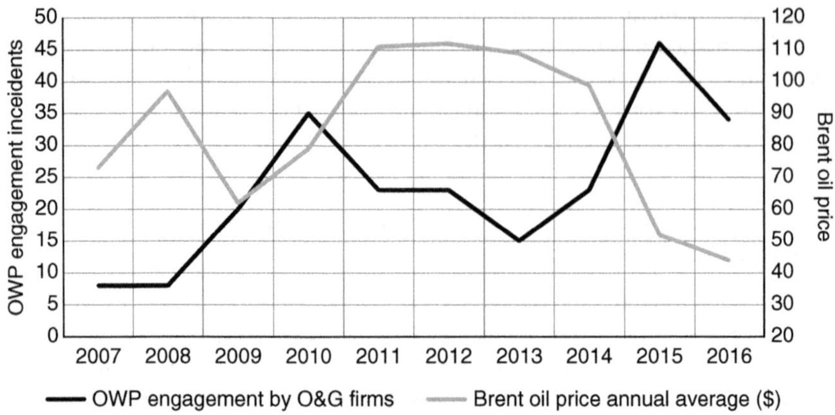

Figure 13.1 Oil price (nominal USD) and engagement of Norwegian O&G industry in offshore wind.

Source: (Mäkitie *et al.*, 2017).

engagement in diversification has followed an inverse pattern with the Brent oil price. The behaviour is therefore somewhat detached from the pattern of continuously growing demand in the European OWP market. In order to advance our understanding of such intermittency in diversification, we investigate five lead diversifier companies in greater detail.

Study of resource redeployment in firms in O&G industry

Statoil

As an O&G company and an operator, Statoil specializes in the production of oil and gas in collaboration with a network of suppliers, and manages technologically complex offshore operations in challenging natural conditions. Its behaviour often signifies the direction of industry activity since many suppliers in the industry provide specific services and products that support the development of Statoil's new field development projects and the operation of existing fields.

Besides its O&G operations, Statoil (since 2018 known as Equinor) has been involved in several renewable energy projects since 2000, including hydrogen and carbon management projects. Statoil's first OWP activities were related to the 'Hywind project', a floating OWP turbine appropriate for deep-water conditions along coastlines unsuitable for conventional bottom-fixed OWP turbines, for example in North America and Japan. Statoil has redeployed technical and operational competences from O&G to the Hywind project such as expertise related to floating foundations which resemble 'spar platforms' (see Chapter 9) utilized in the O&G industry. Moreover, Statoil has redeployed its existing offshore supply-chain connections, its R&D capacity and managerial capabilities (Statoil, 2017). The Hywind project was launched before the financial crisis in 2008 which, combined with the low oil price, had a negative impact on Statoil in 2009 with almost a 30 per cent drop in overall revenue from 2008. Despite the challenges in the primary market, the Hywind demonstration project was introduced during the same year.

Statoil experienced growth again in 2010. The company continued its activities in OWP, this time as an operator and an investor rather than solely as a technology developer. This is closer to its usual role in the O&G market, and has enabled Statoil to use its project execution capabilities in developing large offshore projects. In 2009, Statoil invested in Sheringham Shoal offshore wind park in the UK together with Statkraft, a leading Norwegian producer of electric power. In 2010, Statoil entered into a consortium to develop the world's biggest OWP park so far, Dogger Bank in the UK, and in 2012 the company acquired a majority share in the Dudgeon project. This period was characterized by high oil prices, but was also a period of lower profitability for Statoil.

Lower profitability resulted in the establishment of a cost-reduction programme but did not lead to disengagement from OWP. Despite the drop in oil price in 2014, Statoil announced in 2015 the sanctioning of the first floating wind power farm in the world, Hywind Scotland, with five turbines. The next year the company acquired 50 per cent of the Arkona wind farm, and established Statoil Energy Ventures, an investment fund for renewable energy solutions.

Seen overall, Statoil has diversified to OWP by investing in technology development of floating turbines and in large-scale OWP farms. In floating wind, the company has redeployed scale-free resources originating from O&G such as knowledge of floating technologies, and in OWP parks, its experience, for instance, in operating and developing large offshore projects. Statoil has also redeployed non-scale-free resources such as managerial attention and R&D capacity to OWP, but as OWP has represented only a minor part of Statoil's overall activities, opportunity costs have remained relatively marginal. Indeed, as indicated by the consistent engagement in OWP during 2007–2016, OWP was not abandoned even during the O&G boom of 2011–2014. This suggests that Statoil redeployed resources in OWP due to the growing absolute demand in that market, suffered only minor opportunity costs from this engagement, and had a long-term strategy of diversifying in OWP.

Kværner

Kværner is an established Norwegian industrial company with a long history in Norwegian manufacturing. The company was established in 1853 as a producer of metal products and machines for agriculture, hydropower, and pulp and paper industries, and later as a producer of steel ships and offshore constructions. Through several acquisitions, mergers and de-mergers, Kværner currently describes itself as an EPC (engineering, procurement and construction) specialist that delivers complete offshore platforms and onshore production plants to the global O&G industry.

The steel jacket part of Kvaerner in Verdal (previously known as Aker Verdal), located in Central Norway, entered into a contract to construct 12 steel jackets for a German offshore wind park, Alpha Ventus, in 2007. Additionally, together with other companies in the region, Kværner developed R&D projects and was involved in developing a regional cluster focusing on wind energy in Verdal. Following the financial crisis of 2008, Verdal's business unit was struggling to attract contracts and had to lay off employees. In the aftermath of this crisis, Kvaerner entered into a new EPC-agreement in 2011 to deliver 48 steel jackets and a converter platform in the Nordsee Ost OWP farm. OWP steel jackets resemble similar structures in offshore O&G which enabled Kværner to redeploy engineering capabilities and their production capacity from the O&G market, but also their existing infrastructure in Verdal such as harbours and equipment (Steen and Karlsen, 2014). Moreover, in 2011 Kværner announced R&D efforts in developing OWP jacket foundations and a new converter platform concept for the OWP market.

However, the Nordsee Ost project quickly encountered financial problems, and attempts were made to recover losses caused by additional costs and schedule issues due to changes in the scope of work. In 2012, the former CEO of the Jacket Business was released from his contract, and in 2013 Kværner delivered the last of the jackets and declared its complete exit from the OWP market. While the problematic experiences in OWP played a role in this decision, in the meantime the O&G market had also recovered, and Kværner received considerable new orders from the O&G market and was growing rapidly. The operating revenue of the 'upstream segment' of Kværner, to which the Verdal unit also belongs, rose by almost 80 per cent between 2012 and 2014. However, after the collapse of oil prices in 2014, such strong growth turned into rapid downsizing and nearly halved revenue from 2014 to 2016.

In sum, the possibility to redeploy non-scale-free resources such as production capacity and yards, harbours and other infrastructure (crucial in the production of heavy steel constructions) seems to have been central in Kværner's decision to diversify to OWP (Steen and Karlsen, 2014). Even though scale-free resources such as engineering and production knowledge of steel jackets were also redeployed, the slump in the O&G industry during 2008–2010 created favourable relative demand in OWP compared to O&G, creating incentives for diversification. However, when this diversification attempt ran into trouble, and the relative demand in O&G returned around 2012–2013, Kværner exited the OWP market. This exit was likely easier due to limited sunk costs in OWP because while Kværner had invested in some R&D projects, its investments in OWP were limited due to the resource redeployment of especially non-scale free resources in this diversification effort.

Aibel

Aibel is a company specialized in EPC of offshore projects. Formerly being part of ABB, a major industrial technology company, Aibel was established as an independent company in 2007. Headquartered in Stavanger, south-west Norway, it has several facilities in different parts of the country.

In the aftermath of the financial crisis and subsequent downturn period in the firm, in 2010 Aibel publicly stated itself to be looking into the OWP market. More precisely, the firm hoped to achieve a foothold in building converter platforms. These constructions are similar to platforms which Aibel had previously designed and built in the O&G market. In 2011, together with ABB, Aibel received a major contract to supply such a platform for the DolWin2 OWP farm in Germany. Aibel was to design and build the submersible platform, while ABB would provide the cable and conversion equipment for the converter. Aibel could this way redeploy their experiences from supplying semi-submersible floating platforms in the O&G market. The company started building the platform in 2012 which was finally delivered and installed in 2015.

However, Aibel did not appear to follow up on this early success in OWP during 2012–2013, and no indications of active participation in competition for further projects in the OWP market were observable. At the same time, together with the O&G market, Aibel experienced strong growth, and in 2012 the management reported the intention of hiring approximately 900 new employees. Moreover, the sales revenue of the company nearly doubled in three years from 2010 until 2013. In August 2012, the management also stated that they considered OWP as a niche market supplementing the O&G market while the latter would remain as their core business area.

However, by 2014, the rapid growth had turned into another downturn and the company was making losses and laying off employees. Aibel then returned to the OWP market and in July 2014 it announced a small engineering and management contract of sub-sea foundations for Statoil's Hywind project in Scotland, and an additional small contract in 2016 regarding procurement in the same project. In 2016, the company also announced continuing their collaboration with ABB through a strategic partnership of supplying converter platforms in the OWP market, obviously trying to attract more projects similar to DolWin2.

As an offshore EPC-company, Aibel was able to redeploy its non-scale-free resources related to the design, production and management of offshore projects as well as more general scale-free offshore technology knowledge, without seemingly having to make major investments in new resources. Characterized by redeployment of non-scale free resources in OWP, their engagement in diversification seems to have followed changes in relative demand between OWP and O&G markets. Aibel was active in OWP during downturn periods in O&G, and focussed on O&G during the boom of 2011–2014. Such rapid entry and exit behaviour was likely enabled by high redeployability of resources and few sunk costs in OWP. However, the case of Aibel also shows that earlier success in the OWP market can be difficult to repeat after several years of smaller engagement and little commitment in the new market.

Siem Offshore

Siem Offshore, established in 2005, is a large offshore service company operating in multiple locations around the world. The firm's main assets are 45 vessels (in 2017) used to service and install offshore structures. The company has mainly supplied the upstream O&G industry but recently also established itself in OWP.

The activities of Siem Offshore in OWP commenced in 2011 when they purchased a German company which was renamed as Siem Offshore Contractors. This subsidiary company of Siem Offshore specialized in installing, repairing and maintaining submarine power cables needed on OWP farms. As stated in their 2011 annual report, this enabled the parent company to combine 'the marine operating capacities of Siem Offshore with the engineering capabilities and project execution expertise of Siem Offshore Contractors'. The efforts in entering the OWP market carried first fruit the following year when the company announced contracts at the Amrumbank West and Innogy Nordsee OWP farms. Siem Offshore Contractors also invested in a new OWP-purpose vessel from the maritime company Ulstein. In 2013, the company announced its intention to hire more people on its OWP operations and was actively tendering for more projects. Consequently, it was also awarded another contract associated with the Baltic 2 farm. That year, cable installation activities had amounted to 6 per cent of the total operating revenue of Siem Offshore, increasing in 2014 to 21 per cent as the firm continued to win more contracts.

By 2015, the latest downturn in the O&G industry had hit hard, and Siem Offshore reported heavy losses. However, the OWP activities were growing and becoming increasingly important for the parent company. By 2016, OWP activities already amounted to 41 per cent of the operating revenue and Siem Offshore Contractors produced solid profits. However, the O&G division was struggling due to lack of contracts. Siem Offshore Contractors was awarded four more OWP contracts in 2016, and the parent company has transferred vessels and other resources from the O&G market to meet the demand in OWP operations.

Overall, Siem Offshore diversified by investing in OWP-specific resources through acquisition of a specialized OWP company and commissioning a new purpose-built vessel. This way they could redeploy their existing business and technical knowledge regarding maritime operations gained in the O&G market. Through these sunk costs the firm continued to expand their operations in OWP even during the O&G boom, suggesting a response to the absolute demand in the OWP market with long-term commitment to the market. This strategy of securing a foothold in the emerging market by investing in specialized resources and complementing these with redeployable O&G assets enabled the company to develop 'another leg to stand on', and thus mitigated some of the damage caused by the diminished demand in the O&G market.

Ulstein

Ulstein is a large maritime company involved in ship design, construction of vessels, and maritime system and power solutions. Founded in 1917, the company is headquartered in Ulsteinvik in western Norway.

Ulstein publicly stated its entry to the OWP market in 2011. The company had experienced losses in recent years, and by 2011 the operating revenue of the company had declined by 43 per cent since 2009. In the same year, Ulstein announced plans to develop a new type of OWP installation vessel together with NorWind Installer. However, this venture was followed by only a few OWP activities during 2012 and 2013. The O&G market was nevertheless reviving and the company was slowly but steadily growing. In 2012, Siem Offshore ordered the above-mentioned OWP installation support vessel. This vessel used the X-BOW design of Ulstein, which is particularly useful in offshore operations in the O&G market of the North Sea with difficult weather conditions. The X-BOW design is an inverted bow (i.e. front of the ship), a design concept which decreases vertical motioning caused by waves, thus increasing power efficiency and safety on the ship-deck.

In 2014, Ulstein reported difficult market conditions in the O&G market, and even though they emphasized considering O&G as their key market, the firm declared the intention to look more actively at other markets such as offshore renewables. Specifically, this diversification strategy took place by introducing the new X-STERN concept the same year, which used the same idea as in X-BOW, but this time in the stern (i.e. rear of the ship). This design is especially focussed on OWP service vessels since it makes it easier to approach OWP turbines, and allows technicians to access the turbine in severe weather conditions. In the following years, the difficulties in the O&G sector increased, and the X-STERN concept was in the forefront of the efforts to secure contracts in the OWP market. The first two OWP vessel orders were made by the Bernhard Schulte company. The change in demand between different markets was most visible in the order portfolio of Ulstein yard which hitherto had mostly constructed O&G vessels. In August 2017, three of six vessels on order were designed for OWP and only one to O&G operations.

Ulstein has redeployed scale-free resources to OWP such as knowledge related to ship design (X-BOW and X-STERN). Moreover, non-scale-free resources, e.g. in terms of production capacity, has been utilized. Despite intentions of entering the OWP market in 2011, few such activities took place during the O&G boom. It therefore seems that the relative demand favouring O&G delayed this diversification and as there were few sunk costs in specialized OWP resources, the non-scale-free production capacity and managerial attention was focused in O&G instead. This situation changed after the downturn in O&G in 2014, and the company has recently become more active in OWP.

As can be seen in the above descriptions, the diversification efforts in our case study firms have differed in their nature and timing. Table 13.2 compares

Table 13.2 Comparison of companies

Name	*Ulstein*	*Siem Offshore*	*Aibel*	*Kværner*	*Statoil*
Value chain position	Ship design and building	Offshore services	Engineering, procurement and construction (EPC)	EPC	O&G and OWP operator
Scalability of redeployed resources	Both scale free and non-scale-free	Mostly non-scale-free	Mostly non-scale-free	Mostly non-scale-free	Both scale free and non-scale-free
Redeployed firm resources	Scale-free: design knowledge, design standards (e.g. X-BOW) Non-scale-free: managerial attention, building capacity	Non-scale-free: Vessels and staff	Scale-free: engineering knowledge Non-scale-free: engineering staff, facilities, managerial attention	Scale-free: engineering knowledge Non-scale-free: engineering and building staff, facilities, managerial attention	Scale-free: technological knowledge, managerial capability Non-scale-free: managerial attention, R&D capacity, financial resources
Specialized OWP investments	Limited: R&D (e.g. X-STERN concept)	Yes: acquired a specialized OWP company and an OWP vessel	Limited	Limited: R&D projects and cluster building	Yes: OWP parks and R&D

Market demand at entry to OWP	2011: Good absolute demand in both O&G and OWP	2011: Good absolute demand in both O&G and OWP	2010–2011: Relative demand favouring OWP market, firm struggling to find O&G contracts	2007–2011: Relative demand favouring OWP, firm struggling to attract contracts in O&G	2007: Good absolute demand in both O&G market and in OWP
Reaction to changes in relative demand in markets	2012–2013: Focus on O&G during boom 2014–2016: Renewed interest in OWP when relative demand returned to OWPs favour	2015–2016: Redeployment of vessels and crew in OWP when relative demand favoured OWP after O&G downturn	2012–2014: No new OWP activities during O&G boom 2014–2016: More OWP activities when relative demand returned to favour OWP	2012–2013: After economic difficulties in an OWP project, firm exited OWP and focused solely on O&G after relative demand returned in its favour	No visible reaction, consistent engagement in OWP also through the O&G boom

the diversification behaviour of the five firms in terms of resource redeployment, new investment and development of the market. We further discuss our findings in the following section.

Discussion and conclusions

In order to investigate the oil and gas industry's intermittent engagement in diversification in offshore wind power, we studied resource redeployment between O&G and OWP (Sakhartov and Folta, 2014, 2015). We divided redeployed resources into two categories: scale-free resources, which can be utilized simultaneously in multiple markets (e.g. knowledge, standards and experience), and non-scale-free resources which cannot be applied in multiple use without limitations (e.g. human resources, facilities and managerial attention) (Levinthal and Wu, 2010). Moreover, we sought information about the extent of investments by firms in specialized OWP resources. We analyzed patterns of entering and exiting the diversified market in the light of changes in absolute demand (i.e. the degree of business opportunities in a market) and relative demand (i.e. relative attractiveness between the two markets in terms of business opportunities) (Lieberman *et al.*, 2017; Wu, 2013).

From the five companies we have studied, Statoil and Siem Offshore had the most consistent level of engagement in the OWP market which lasted throughout the O&G boom period. Common to both companies was that they had invested in OWP-specific resources early in their process of diversification: Statoil in the development of floating wind power technology and in OWP farms, and Siem Offshore in acquiring an OWP company and commissioning a vessel. Statoil redeployed both scale-free resources (such as technological and managerial knowledge related to offshore conditions) and non-scale-free resources (such as financial resources, R&D and managerial attention). However, due to the size of the company, the latter is not likely to result in a major opportunity cost for the company. For Siem Offshore, investment in a specialized OWP segment in the company enabled growth in OWP, also during an O&G boom, without major opportunity costs.

The EPC companies, Kværner and Aibel, had more intermittent diversification patterns in OWP. Both companies redeployed non-scale-free resources to OWP, presumably seeking to meet the lack of demand in the O&G market, and redeployed these resources back to O&G when that market picked up. Due to problematic experiences, among other reasons, Kværner decided to exit OWP, but Aibel has sought to engage in OWP again during the latest decline in the O&G industry.

On the other hand, the maritime company Ulstein redeployed both scale-free (e.g. vessel design concepts) and non-scale-free (e.g. production capacity) resources together with some investments in specialized OWP resources. The company seemed to have had a similar on-off pattern in engagement in diversification as Aibel, but with slightly better success in acquiring contracts in the OWP market after the latest decline in the O&G market. Therefore, as a versatile company with different types of redeployed resources, it had characteristics of both consistent and intermittent engagement in diversification.

The above results support the arguments of Wu and Levinthal (Levinthal and Wu, 2010; Wu, 2013) and Lieberman and colleagues (Lieberman *et al.*, 2017) in suggesting that types of redeployed resources (through existence or absence of opportunity costs) and sunk costs have a role in shaping the engagement in diversification to related markets. Our analysis suggests that firms which used OWP to merely redeploy non-scale-free resources during periods of diminished market opportunity in O&G, combined with limited investments in specialized OWP resources, had the most intermittent engagement in diversification. In comparison, companies that combined (scale-free) resource redeployment with significant investments in OWP-specific resources have been more able to grow OWP business as an integral part of the company's activities.

Such results have implications for industrial policy in countries which seek to utilize the resources of existing resource-based industries in developing new industries (Mahroum and Al-Saleh, 2017). Our results suggest that industrial policy which seeks to encourage diversification of firms should distinguish between scale-free and non-scale-free resource redeployment. The latter is more likely to be associated with temporary changes in the relative attractiveness of alternative markets, while the former has more chances for sustained engagement in diversification. Hence, in order to allow new industries to grow through the sustained diversification efforts of companies, industrial policy should support the redeployment of scale-free resources. Moreover, policy measures should also support investments in new specialized resources in the diversified market. Such investments will assist firms in sustaining their competitiveness in the new market, and solidifying their commitment to diversification through sunk costs. This latter point is important due to the potential munificence of the O&G market, i.e. the attractive market opportunities and high profits during boom periods in comparison to most other markets.

We conclude that our study has drawn attention to three little discussed perspectives in diversification of the O&G industry in other markets: 1) scalability of redeployed firm resources and opportunity costs, 2) investments and sunk costs and 3) relative changes in market demand. We argue that such perspectives can be helpful in understanding the observed intermittency in the O&G industry's engagement in diversification (cf. Mäkitie *et al.*, 2017). However, our study also has limitations which offer opportunities for further study. While we see our empirical study as useful in highlighting the three above-mentioned perspectives on diversification, our study is not meant as a full assessment of all underlying factors influencing the diversification efforts of the O&G industry. Also, as our study has been limited in studying just a handful of companies and has used only secondary data sources, we encourage further studies to investigate resource redeployment in the diversification of natural resource supplier industries both at the firm-level as well as at the industry-level.

References

Barney, J. (1991). Firm resources and sustained competitive advantage. *Journal of Management*, 17(1), 99.

Dixit, A. (1989). Entry and Exit Decisions under Uncertainty. *Journal of Political Economy*, 97(3), 620–638.

Hansen, G. H., and Steen, M. (2015). Offshore oil and gas firms' involvement in offshore wind: Technological frames and undercurrents. *Environmental Innovation and Societal Transitions*, 17, 1–14.

Helfat, C. E., and Eisenhardt, K. M. (2004). Inter-temporal economies of scope, organizational modularity, and the dynamics of diversification. *Strategic Management Journal*, 25(13), 1217–1232.

Helfat, C. E., and Lieberman, M. B. (2002). The birth of capabilities: market entry and the importance of pre-history. *Industrial and Corporate Change*, 11(4), 725–760.

Kern, F., Smith, A., Shaw, C., Raven, R., and Verhees, B. (2014). From laggard to leader: Explaining offshore wind developments in the UK. *Energy Policy* 69, 635–646.

Lee, G. K., and Lieberman, M. B. (2010). Acquisition vs. internal development as modes of market entry. *Strategic Management Journal*, 31(2), 140–158.

Levinthal, D. A., and Wu, B. (2010). Opportunity costs and non-scale free capabilities: profit maximization, corporate scope, and profit margins. *Strategic Management Journal*, 31(7), 780–801.

Lieberman, M. B., Lee, G. K., and Folta, T. B. (2017). Entry, exit, and the potential for resource redeployment. *Strategic Management Journal*, 38(3), 526–544.

Mahroum, S., and Al-Saleh, Y. (eds.). (2017). *Economic Diversification Policies in Natural Resource Rich Economies*: Abingdon: Routledge.

Mäkitie, T., Andersen, A. D., Hanson, J., Normann, H. E. and Thune, T. M. (2018). Established sectors expediting clean technology industries? The Norwegian oil and gas sector's influence on offshore wind power. *Journal of Cleaner Production*, 177, 813–823.

Mäkitie, T., Normann, H. E., Sraml Gonzalez, J. S. and Thune, T. M. (2017). *The green fling: The role of market volatility in incumbent industries' engagement in sustainability transitions*. TIK Working Papers on Innovation Studies 20180524, Centre for Technology, Innovation and Culture, University of Oslo.

Markides, C. C. and Williamson, P. J. (1994). Related Diversification, Core Competencies and Corporate Performance. *Strategic Management Journal* 15, 149–165.

Montgomery, C. A. and Hariharan, S. (1991). Diversified expansion by large established firms. *Journal of Economic Behavior & Organization*, 15(1), 71–89.

Normann, H. E. (2015). The role of politics in sustainable transitions: The rise and decline of offshore wind in Norway. *Environmental Innovation and Societal Transitions* 15, 180–193.

Normann, H. E. and Hanson, J. (2017). The role of domestic markets in international technological innovation systems. *Industry and Innovation*, 1–23.

Penrose, E. (1959). *The theory of the growth of the firm*. Oxford: Blackwell.

Pisano, G. (2016). Towards a Prescriptive Theory of Dynamic Capabilities: Connecting Strategic Choice, Learning, and Competition. *Harvard Business School Working Paper*, 16–146.

Prahalad, C. K. and Bettis, R. A. (1986). The dominant logic: A new linkage between diversity and performance. *Strategic Management Journal*, 7(6), 485–501.

Rumelt, R. P. (1974). *Strategy, structure, and economic performance*. Boston, Mass: Harvard University Press.

Sakhartov, A. V., and Folta, T. B. (2014). Resource relatedness, redeployability, and firm value. *Strategic Management Journal*, 35(12), 1781–1797.

Sakhartov, A. V., and Folta, T. B. (2015). Getting beyond relatedness as a driver of corporate value. *Strategic Management Journal*, 36(13), 1939–1959.

Steen, M. and Karlsen, A. (2014). Path creation in a single-industry town: The case of Verdal and Windcluster Mid-Norway. *Norsk Geografisk Tidsskrift – Norwegian Journal of Geography*, 68(2), 133–143.

Tanriverdi, H. and Venkatraman, N. (2005). Knowledge relatedness and the performance of multibusiness firms. *Strategic Management Journal*, 26(2), 97–119.

Teece, D. J. (1982). Towards an economic theory of the multiproduct firm. *Journal of Economic Behavior & Organization*, 3(1), 39–63.

Wieczorek, A. J., Hekkert, M. P., Coenen, L. and Harmsen, R. (2015). Broadening the national focus in technological innovation system analysis: The case of offshore wind. *Environmental Innovation and Societal Transitions*, 14, 128–148.

Wu, B. (2013). Opportunity costs, industry dynamics, and corporate diversification: Evidence from the cardiovascular medical device industry, 1976–2004. *Strategic Management Journal*, 34(11), 1265–1287.

Additional sources

Statoil. (2017). Making the most of our advantages. Retrieved 22.02.2018 www.statoil.com/en/what-we-do/hywind-where-the-wind-takes-us/making-the-most-of-our-advantages-.html.

Part IV

14 The resource endowment challenge

Extending the value chain

Øystein Noreng

Introduction – Norwegian milestones

Norway's petroleum history commenced in 1962 when a group of oil companies sought permission to explore for oil on the Norwegian continental shelf. Norway's response was to wait and study the issue. The decision built on a protectionist tradition in national resource policy as in fisheries and hydroelectricity since the early twentieth century. Political consensus was that any oil industry on the continental shelf should be under Norwegian control, with strong state participation if needed. Indeed, Norway is an example of an entrepreneurial state with a broader and longer view than market forces, and able to make rational strategic choices (Mazzucato, 2015). At that time, the international oil industry was dominated by a cartel of integrated oil companies that also controlled technology through the supply and service industries. In terms of resource nationalism, Norway was well ahead of most oil producing countries.

The Petroleum Act of 1965 provided a legal framework, defining offshore natural resources as being the property of the state which was free to impose special measures and taxes. The choice was for a concessionary system based on discretionary allocations, not auctions. State participation of at least 35 per cent in all new concessions was introduced in 1969, and the first oil discovery was made in the same year.

In 1972, Statoil and the Norwegian Petroleum Directorate were established, followed by provisions for the use of Norwegian goods and services and technology transfer as well as provision for a state majority in all new concessions. In 1973, the Statfjord discovery was made. Soon after, oil prices quadrupled. High oil prices triggered a policy response in 1975 with Special Petroleum Taxation, aimed at taking a large share of the accrued economic rent. In 1979, the Troll gas field was discovered.

A dominant position in Norway's petroleum industry together with high oil prices aroused concern that Statoil's handling of its cash-flow would have macroeconomic repercussions. In 1983, the State Direct Financial Involvement (SDFI) was introduced as direct participation in which the state took a given share of costs and gross income, curbing Statoil's cash-flow.

The decline in oil prices in 1986 resulted in reduced interest in exploration by the oil industry. Petroleum taxes were reduced accompanied by persistent efforts to cut costs. In 1992, the Petroleum Fund was established in order to protect the domestic economy from oil price volatility.

During the 1990s, the bulk of investment was made by Statoil and the SDFI. In 1999, the industry became consolidated whereby three major national oil companies were reduced to two. In 2001, Statoil was partly privatised, the company subsequently opting for international expansion. At this time, with low oil prices and low activity, the Norwegian oil industry was widely regarded as a sunset industry. A mature resource base needs rising oil prices and/or lower costs. The oil price rise in the wake of the Iraq War was highly beneficial to Norway.

The 2004 petroleum tax change with exploration compensation for companies not yet in a tax-paying position caused an influx of newcomers – smaller, independent companies with fresh ideas. The outcome was increasing activity, more exploration, more finds and more development, leading to higher output. During the expansion period up to 2014, costs increased due to bottlenecks and strained supply chains. Drilling accounts for at least half of the total costs in the Norwegian oil industry and rising rig rates triggered a general cost increase. Since 2015 consolidation and cost cuts are on the agenda. Lower activity has caused costs to decline.

Local content – a global issue

The challenge is to establish real economic linkages between the oil and gas sector and the rest of the economy. Many oil exporters have the resources and financial endowment, but they lack skills. The development of human capital is a condition for the successful development of a local petroleum cluster. Experiences vary. Norway is a leading exporter of petroleum technology. In many cases, employment seems at least as important as research and development.

In the case of Norway, an appropriate question is the extent to which the various protectionist policies in relation to petroleum concerning local content, transfer of knowledge and industrial employment did not strengthen the dependence on the petroleum industry. This represented a bout of 'Dutch disease' in disguise, a resource curse intensified by institutions. Indeed, the local content policy had made parts of Norwegian manufacturing exceedingly dependent on the local offshore market. The knowledge transfer policy had concentrated the country's research efforts on petroleum-related matters. Significantly, as these policies of microeconomic intervention were abandoned in the mid-1990s, there was a general understanding that they had contributed to high costs in the petroleum industry as in other parts of the economy. Low oil prices resulted in an immediate need for structural change.

Infant industry protection

From the early 1970s to the mid-1990s, Norwegian petroleum policy represents a fairly systematic case of infant industry protection. Traditionally, infant industries have been protected by tariffs. For a generation, Norwegian government regulations may have been more effective as tools of microeconomic intervention than tariffs. Historically, oil companies competed for licences by virtue of their technical competence as well as their local contribution. Technological and managerial prowess are still key criteria for licence allocation (see Chapter 9).

The argument in favour of protecting an infant industry is the need for temporary protection of a new activity that in the short run would be unable to survive competition with established foreign suppliers, but which has the potential to become competitive, the bottom line being that the need for protection will diminish over time. Infant industry protection is a transient arrangement where in the short run social costs are supposedly offset by social gains in the long term. This would require the increase in future national income to exceed what it would have been if the infant industry had not been protected. The argument is that infant industry protection is warranted for small new firms that have little chance of competing with established firms which have been in business longer and over time have been able to improve their efficiency. They are likely to have developed competitive advantages, exploited economies of scale and gained superior knowledge of markets and technology.

This argument is mostly used in conjunction with establishing new industries in developing countries, to justify protection against established competitors in industrial countries. The argument is also used in conjunction with alleged imperfect competition. This was the case with the Norwegian oil industry around 1970. The policy choice was to promote local content in a way that in a short time established a large captive market for petroleum-related goods and services. The very success of the infant industry protection policy may inadvertently have enhanced the oil risk exposure of the Norwegian economy as the new industries were also dependent on oil prices. Indirectly, this might be seen as strengthening the resource curse.

The established hypothesis is that the resource curse with a high-risk exposure is practically unavoidable for countries whose governments have access to economic rent since after adjusting for other relevant factors, their expenditure causes significantly higher prices in resource-dominated economies, compromising the competitiveness of the non-resource-based parts of the economy (Sachs and Warner, 2001). The 'Dutch disease' is the classic example. It means that the discovery and exploitation of natural resources deindustrialise a nation's economy. As costs rise, the currency appreciates making manufactured goods less competitive while imports increase and non-resource exports decrease. The outcome is high-risk exposure, instability and precarious public finances and employment. A contrary hypothesis is that countries with large natural resource endowments constitute both growth losers and growth winners, the main reason being differences in the quality of institutions (Mehlum *et al.*, 2005). Apparently, this should apply to Norway together with, for example, Australia and Canada.

The value for society of supporting and protecting infant industries by investing in human capital and limiting foreign competition has been an issue among economists for generations; opinion continues to alternate between reliance on market forces and government intervention. In the case of Norway, the question is whether the use of local goods and services and the training of nationals would have evolved in any case on the basis of competitive locations and skills, but possibly at a lower cost through the market than through state intervention. The question is also whether market forces would have produced comparable results just as quickly.

Procurement, local content and contracting

The Norwegian success in achieving high local content is largely due to government policies which encouraged partnerships between foreign and domestic companies, and demanded that foreign companies supported development of domestic R&D capabilities.

At the contract award stage, operators were obliged to inform the Ministry of their evaluation with the recommended supplier, together with details of price, country of origin and Norwegian content. The Norwegian content was calculated as value-added in Norway, both in manpower and monetary values, regardless of the ownership of firms. The Ministry did not specify content nor supplier. In practice, the Ministry set targets for the share of the value-added to be realised by Norwegian suppliers, but left the oil companies to choose how targets were to be reached, leaving the oil company management with the discretion to decide what parts and services should be procured locally, and from which suppliers.

The policy was first applied in the third licensing round in 1973. Field development operators had to present a plan to the Ministry for all tenders above $150,000 at the current exchange rate. Prior to tender invitations, the operator had to announce the tender schedule and the companies to be invited. The Ministry's role was to ensure that qualified Norwegian companies were included on the bidder's list.

At the contract award stage, the operator was to inform the Ministry of their evaluation with the recommended supplier, price, country of origin and Norwegian content. The role of the Ministry was to ensure that a Norwegian bidder was awarded the contract when competitive in terms of price, quality, delivery time and service.

The emphasis placed on local content made it essential for all oil companies to use Norwegian goods and services. The Ministry used Norwegian content as one of the evaluation criteria when evaluating companies competing for new acreage. The Ministry's policy was to be transparent and predictable regarding enforcement of the procurement policy, publicising who used what and from which local suppliers. Moreover, as an observer in all licence groups, the Ministry secured insight into all the operators' contracting activities.

The Ministry abstained from intervening directly in the procurement process; neither were the oil companies instructed what to buy from which suppliers. It had no competence to evaluate the specific oil company's requirements nor the supplier's qualifications. Abstaining from direct intervention also insulated the Ministry from political pressure by specific firms and regional authorities to secure contracts, profits and jobs. The effect of the general target was that in order to comply with the local content requirement measured by value-added, oil companies generally preferred to procure high-value parts and equipment locally, responding to high Norwegian wages, and also contributing to technology development in Norway, importing less sophisticated parts from abroad.

The conclusion of the Free Trade Agreement in the European Economic Area in 1994 between Norway and the European Union meant that the procurement policy had to be discontinued. In hindsight, by 1994 the procurement policy had served its purpose and the Norwegian oil service industry was ready for exposure to international competition.

Transferring skills

Already around 1970, realising the petroleum potential and skills gap, the Norwegian government actively promoted higher education in petroleum geology and technology (Cleary, 2016). From the very beginning, Statoil sent young employees to training establishments abroad. Statoil also hired highly skilled US personnel as an alternative to depending on US companies. On the industry side, the requirement to transfer competence and to cooperate in the development of new technology was introduced in the third licensing round in 1973. The US company Mobil Oil was operator of the Statfjord field, but was required to systematically train Statoil personnel to take over operations tasks.

The licence terms for the international oil companies made it mandatory to transfer skills and competence to the Norwegian companies. Personnel from the Norwegian oil companies participated initially in the oil majors' training courses and received on-the-job training as part of their overseas operations. The oil majors recruited young Norwegian engineers and trained them overseas for a significant period before they returned home.

The transfer of technology and cooperation in research and development has been among the most successful aspects of Norwegian petroleum policy. By compelling oil companies to transfer competence and to cooperate in the development of new technology, Norway could assume the role of a leader in international petroleum development. Within a relatively short time, Norwegian competence and technology suited to Norwegian conditions was developed. Competence strengthened Norway's bargaining position in the international oil industry while technology development led to significant cost reductions and an ensuing expansion of the resource base. The background was the strong research and development effort and cooperation between oil companies, the supply industry and research institutions.

The fourth licensing round in 1979 introduced provisions for technology development. Cooperation between foreign oil companies and Norwegian research institutions resulted in the former contributing funding, insight and expertise in the development of technology in Norway.

Three types of agreement between foreign oil companies and Norwegian research institutions were applicable:

1 Fifty per cent agreements required operators to conduct at least 50 per cent of the research and development needed to develop a prospect in Norway at Norwegian institutions.
2 Agreements used in the fourth and fifth licensing rounds required operators to conduct a specified research effort in advance of new licensing.
3 Goodwill agreements, where the oil companies were to conduct as much petroleum-related research and development as possible in Norway, without any advance commitment as to the sum or volume of the effort. Initially, the research and development Goodwill programmes were based on dialogue with the industry and were developed to support its needs and priorities.

The value of the technology cooperation should be seen in the light of the lead times for technology breakthroughs which is usually ten to 15 years. The results were that the Norwegian oil industry experienced major advances in drilling and subsea technologies, the application of information technology, and hence, cost reduction.

The ambition

In hindsight, the policy of knowledge transfer expressed an ambition to use a natural resource-based, extractive and capital-intensive industry to strengthen the basis of a knowledge-based economy. In terms of economic dynamics, the oil industry covers qualitatively different activities from the point of view of value-added creation:

1 The core oil industry is an extractive activity whose value-added contains a large share of economic rent.
2 The supply industry is a manufacturing activity or routine service whose value-added is essentially based on labour productivity and management quality.
3 Research and development is an intellectual activity whose value-added is essentially based on knowledge and inventiveness.

The challenge for Norway has been, and still is, to compensate the depletion of a finite resource with the creation of durable assets. Human capital – knowledge

– is considered more durable than real capital in manufacturing assets, as in the longer run investment in education outperforms practically any other investment. From this perspective, Norway's oil policy has contained a knowledge strategy providing for a long-term income base, parallel to the investment fund (see Chapters 2 and 7).

To extract or not to extract? That is the question

As the landowner, the state has distinctive responsibilities and interests, different to those of private owners and investors. Moreover, the landowner-state potentially controls huge wealth. First, the state has an overall responsibility for an entire national economy, not only for one sector, meaning the need to balance different and, at times, conflicting concerns. Second, the state has a responsibility for long-term sustainable development, also taking into account the concern for future generations. This is particularly pertinent in the case of extractive industries where the pitfall is that the depletion of the finite resource base finances consumption, but where the challenge is to use one-time revenues to finance investment in new sources of income, including human capital.

The task cannot be left to the market alone. Markets are generally short-sighted, with a propensity for choosing simple, convenient outcomes, with little concern for the economy as a whole, for social effects or the longer term. Moreover, markets can be imperfect, dominated and governed by a small number of strong actors. Therefore, as the resource owner, the state needs to elaborate a strategy for the purpose of the activity and the use of revenues beyond the status quo. This is the impetus for the entrepreneurial state.

Depletion is the key to resource policy, whether it concerns oil and natural gas, or metals and minerals. One option is to extract the resource as quickly as possible and spend the proceeds without much concern for the future. The alternative is to extract with end-use for the proceeds in mind. This implies a conscious trade-off between consumption and investment.

The critical parameters are not evident. The absorptive capacity and marginal returns on investment are a matter of where the money is invested and in what way. The return on foreign investment is uncertain and should consider financial market risks. Likewise, leaving resources in the ground means exposure to long-term oil price fluctuations.

Oil exporters face the trade-off according to their differing economic conditions. In the 1950s and 1960s, most oil exporters were at a stage of development where human resources were under-utilised. Leaving oil in the ground was considered risky as the oil exporters were competing for funds and technology from the international oil industry.

At that time, Norway – a newcomer to the game – was in a different situation. Full employment and a trade surplus indicated a low absorptive capacity, and a modest marginal return on domestic investment. Indeed, the established industries viewed the emerging oil industry with circumspection, fearing economic overheating, cost inflation and loss of competitiveness. Therefore, at an early stage Norway opted for a moderate rate of extraction *and* a moderate domestic use of oil revenues, implicitly making a strategic choice to invest *both* in foreign assets *and* in oil in the ground. The point was to shelter the domestic economy from unwanted pressures for rapid structural change.

In the late 1960s, in contrast to the United Kingdom, Norway had no domestic oil industry. There were no forces on the political scene that had vested interests in oil, except for ship-owners keen to make money from the emerging oil industry. The Norwegian government was free to set its own agenda and set itself at a distance from the international oil industry.

Entering the value chain

Conventional oil and gas projects usually require heavy capital investment for a limited period, followed by years of low lifting costs that gradually rise over the lifetime of a prospect. Manufacturing equipment requires specialised facilities as well as specialised knowledge. Due to the large domestic oil industry, the United States has been the world leader in the oil supply and service industries, with a small number of American firms dominating the world market, also with oligopolistic features. The size of the home market was decisive.

The large international oil companies developed symbiotic links with the supply and service industry. Two oligopolistic structures developed long-term business relations. At the time, petroleum research and development was essentially in the hands of the oil companies so that petroleum technology was largely their exclusive domain. No oil producing country could operate outside the circuit. The exception was the Soviet Union with a large domestic oil industry and an indigenous supply and service industry, excluded from Western research and development and therefore, by many accounts, of inferior quality.

France, Italy and the United Kingdom developed their own supply and service industries in cooperation with their oil companies, but the British, French and Italian supply and service industries could not match their American counterparts over the spectre of the business. Other oil producers did not have the resources to establish oil supply and service industries, not even Canada. The need for investment funds and technology made the developing oil producing countries hostages of the international oil companies and their supply and service partners.

Around 1970, Norway entered the stage with a different background and ambitions of its own. Although Norway had no previous experience in upstream oil activity, it had a maritime experience second to none. It also had a shipbuilding industry and a supply service industry serving the merchant navy, with a host of companies eager to get into the oil business. Consequently, there was strong backing for protecting national interests. The Norwegian institutional model was already in place by 1972 with strong elements of protectionism and resource nationalism (see Chapter 7).

Remarkably, this took place before the OPEC countries took control of prices and production. The strategic choice was based on the assumption that the resource base was ample and would be able to sustain petroleum activities for decades. For that reason, not only was national control of the petroleum activities recommended, implying exploration, development and extraction, but also entry into input parts of the value chain, the supply of goods and services. The reasoning was simply that it is difficult to control an industry without insight, and that this was best achieved through active participation.

The industrial policy

At the outset of petroleum activity in the early 1960s, Norway had no indigenous oil nor oil services industry, but around 1970 there was consensus that it had to create one with the help of the government since market forces alone were not regarded as capable. By 1972, a procurement policy was in place.

In hindsight, the procurement policy was successful, bringing the local content of investment from almost nothing in the late 1960s to almost 90 per cent 20 years later. The gain was a large domestic oil supplies industry, manufacturing parts, delivering services, and a remarkable technological development to overcome the challenges of a difficult environment as well as job creation. The disadvantages were a local cost pressure and an excessive dependence on petroleum development. Since the protectionist and preferential policies were discontinued in 1994, data on local content are no longer regarded as reliable, but currently the Ministry estimates the share of local content to be about two-thirds.

Today, Norway has a large oil service industry that is increasingly turning its attention to export markets. The procurement policy has also strengthened employment, although the capital-intensive petroleum industry is no great job generator.

Through the knowledge policy, Norway has established itself as a leader in offshore- and especially deep-water petroleum technology. The consistent research effort, with the oil companies contributing competence, funds and personnel, laid the ground for the remarkable technology development that, since the mid-1980s, has reduced unit costs in petroleum development by 3 to 5 per cent annually, and developed services for a world market.

Against this backdrop, Norway has been successful in local content development and job creation. The result is a strong manufacturing base for the oil industry which is relevant to other oil exporters in need of job creation. The cause was targeted government policies, and subsequently the essential role of Statoil, the national oil company, in implementing and continuing government policies. The government policies of Norway during the period 1972–94 are relevant to the present situation in many oil producing countries.

Statoil's role, before and after the 1994 policy changes is directly relevant to other national oil companies, and some of Statoil's methods to initiate and assist new businesses might be adopted by other countries. In Norway, national control of the industry has been essential for realising the profit potential of the entire value chain and implementing a procurement policy giving preference to local suppliers. Similarly, developing a national oil supplies and services industry meant avoiding dependence on foreigners and keeping part of the economic rent at home. Regardless of the protectionist measures, bottlenecks and imperfect competition in the value chain mean that part of the economic resource rent will move to the supply and service contractors.

Remarkably, the very success of the industrial policy in relation to petroleum has been a macroeconomic constraint. At least 60 per cent of offshore investment represents deliveries by Norwegian suppliers (Eika *et al.*, 2010). For maintenance the share is higher. For many years, from 2006 until the oil price slump in 2014–16, these direct demand effects of the petroleum industry were higher than the indirect effects of using petroleum revenues over the budget. At the same time, concerns about overheating the economy incited the current centre-left government to moderate the use of oil money. The petroleum tax system meant that around 90 per cent of the investment risk was carried by the state.

The entrepreneurial state has been less effective in controlling the oil industry that it has created. Whereas listed private firms are under constant supervision by capital markets and pressure to perform, state-owned firms are in a different situation. Norway's Statoil, like other state-owned companies, was established with a capital and resource base endowment from the state, with a mandate to serve comprehensive and long-term national interests, whose value is difficult to quantify, and therefore without any specific requests for returns on investment. The usual pattern is that after a short time state-owned oil companies become independent bodies, not subject to the discipline of capital markets. With an illusory state supervision, they rule themselves and define their own strategies, enjoying more managerial discretion than is the case in corresponding private firms listed on the stock exchange. Dividends make a key difference. Investors in private oil companies usually request high dividends when oil prices and earnings rise, but state owners are less perseverant in this regard, so that the companies can retain more of the cash-flow. The risk is uncritical expansion. The absorption of two other oil companies has given Statoil an unhealthy dominant position that could have been avoided by an alert government.

Lack of effective external control is generally a recipe for mistakes (Olson, 1971). Such is the case with Statoil that in the late 1980s it had a major cost overrun. On several occasions, after scandals of various kinds, the company's senior management was fired.

In hindsight, the partial privatisation of Statoil in 2001 has not been an unequivocally lucky stroke. In the hope that a minority private participation of 30 per cent would discipline the company, it was let loose from government supervision. The senior management benefited from the liberty to opt for a strategy of international expansion. Even though several of the foreign ventures have been successful, Statoil's investment in North American unconventional oil has brought heavy losses. From 2001 until 2016, Statoil invested more upstream outside Norway than within. High profits on Norwegian operations have financed loss-making ventures abroad. During the years of high oil prices, the government's neglect to request correspondingly high dividends left Statoil with a generous cash-flow to spend on foreign investment, much of which had to be written off. The lesson is that the entrepreneurial state has yet to learn how to discipline its enterprises and set limits for their activities. The relationship between the Norwegian government and Statoil is a typical case of the Principal-Agent problem.

The challenge of maturity

Oil provinces go through different phases of a life cycle. The initial phase, the inception – meaning high-risk exploration in virgin territory – is often carried out by smaller companies. Following oil discovery, large oil companies move in with financial and human resources to develop the infrastructure and with big prospects. They usually dominate the stage during the build-up of production through the peak into maturity. As the resource base matures, prospects become smaller and more heterogeneous. The huge integrated energy companies have an evident interest in balancing their oil and gas needs by equity production, but their size and high overhead costs make them most suited to take on large projects. A diverse group of small and medium-sized oil companies that are specialised and have lower overhead costs enter thereafter. These are more able to engage in the development of smaller and marginal prospects. This trend has been evident in the US Gulf of Mexico and the UK continental shelf for decades. With advancing resource maturity this is evident in Norway where the tax changes introduced in 2005 facilitated the entry of newcomers into a mature oil province. Measured in terms of exploration activity and discoveries, this measure has been successful. The newcomers – smaller oil companies – are more flexible and more innovative. They have to offset their limited capital resources by smarter thinking. Their limited resources mean that their research and development is essentially carried out together with the supply and service industry.

The petroleum fund

Since the mid-1990s, Norwegian policy has been to accumulate the oil and natural gas revenues in an investment fund. The intention was to protect the domestic economy from oil price fluctuations and to avoid harmful domestic inflationary pressures due to the expenditure of economic rent. So far, this has been reasonably successful. Norway has managed to combine high employment with low inflation, including through the financial crisis of 2008–9 and the oil price crash of 2014–16.

The policy of saving the oil and gas revenues has been even more successful. The Petroleum Fund, Norway's sovereign wealth fund, had a value of about one trillion US dollars by the summer of 2017, equivalent to two-and-a-half times Norway's gross domestic product. The Fund has diversified investment in bonds, stock and to a lesser extent real estate. From 1992 to 2016 the Norwegian government has transferred seven trillion NOK to the Fund; at the end of 2016 the

Fund's market value was NOK 7.5 trillion, doubling the nominal input value. By comparison, Germany has had a balance of payments surplus of US$2.6 trillion between 2000 and 2016; total foreign assets had a value of US$1.6 trillion by the end 2016. One trillion dollars was lost, mainly during the financial crisis in the United States and Southern Europe. German foreign investment is managed by banks and large industrial corporations, at times with high risk and heavy losses. Capital asset management by the Norwegian Petroleum Fund, under the auspices of the Central Bank, seems to have been more successful. Norwegian portfolio investment has taken a lower risk and yielded a higher return than the bulk of German direct investment in companies.

The Petroleum Fund diversifies the Norwegian economy, using petroleum revenues to create other sources of income. It also represents an extension of the value chain. The amount transferred to the budget is determined by politicians according to advice and convenience.

By 2016, the Fund had become a self-financing money machine. Investment income was higher than the out-take – the transfer to the budget. The oil revenue decline was partly compensated by higher investment income. Even with lower oil prices, the Fund remained in surplus, and Norway's financial wealth continued to grow. Thus, within a few years, the Petroleum Fund had become an investment fund, but with lower annual income and more exposure to financial market risks.

Until the financial crisis, the out-take from the Fund was modest, equivalent to between 2 and 3 per cent of GDP. Facing a sudden decline in oil revenues after 2015, the centre-right government stepped up the transfer from the Fund, reaching 8 per cent of GDP in 2017, an election year. The out-take is still below total income, but exceeds both oil revenues and returns on investment. The ceiling for transfers has been lowered to 3 per cent of the Fund's market value, about the 2017 level. It is still a considerable amount in relation to Norway's GDP.

Ironically, the outcome might be that Norway, in the first instance having avoided the resource curse, in the second instance develops a money curse. The resource curse would emerge indirectly, through a financial endowment built on the proceeds from resource extraction that were not spent instantly, but saved. Having avoided the classical curse of the resource rent, Norway nevertheless seems to retain a rentier economy, with the financial rent gradually substituting a diminishing petroleum rent. The bottom line is that Norwegian politicians will retain a considerable freedom in economic policies. The temptation is to use the money solely to maintain living standards and social services. The challenge is to use the money for a more profound diversification of the economy, by badly-needed infrastructure investment and an increased investment in research and industrial development.

The risk is that the resource curse materialises through a financial endowment. In that case the benefit for the country of keeping oil and natural gas in the ground might exceed the value of financial surpluses, once more raising the question of whether to extract or not to extract. If this was to be the case, the resource curse hypothesis of Sachs and Warner (2001) would be verified. Alternatively, the contrary hypothesis of Mehlum *et al.*, (2005) would seem plausible. However, the institutions were also grabber-friendly, underlined by the early use of anticipated oil revenues in the 1970s to support farmers, ship-owners and manufacturing in general. By the mid-1990s, the oversized oil supply and service industry again indicated that Norway's institutional framework had some grabber-friendly aspects. Since lifting protectionism, the export performance of the oil supply and service industry indicates international competitiveness, also that Norway's institutions are producer-friendly.

The alternative hypothesis put forth by Mehlum *et al.* (2005) seems to assume that political institutions are neutral in relation to their economic base, an assumption hardly supported by the evidence. Large resource revenues make the government more independent of a productive income base and of taxes levied on civil society, usually leading to a consolidation of power, and in developing countries often to autocratic rulers. In many cases, historically high oil prices could reduce the need for income taxes. In established democracies such as the Netherlands and Norway, large resource revenues decouple expenditure from productivity to some extent, broadening the freedom of action in economic policy, making the political system a more tempting battleground for interest groups, and encouraging arbitrariness and rent-seeking through selective measures. Norway is a case in point.

Another question is to what extent the reliance on market forces, with a less interventionist state in petroleum matters and a stronger position for foreign companies, would have produced comparable results in areas that matter to the Norwegian public, namely the integration of the petroleum industry into Norwegian society through jobs, deliveries and national control. This is a complex issue, not limited to monetary costs and benefits, but also to qualitative aspects of Norwegian society that are difficult to quantify, let alone express in monetary terms. In the early 1970s, the emerging petroleum activities and associated policies were the subject of intense political debate in Norway, with the government presenting alternative policies and risks, something which has been described as exemplary for democratic deliberation of resource policy.

Nevertheless, having avoided the 'Dutch disease', the risk is that Norway will develop a 'Norwegian disease', a rentier economy where high income from huge financial assets from the exploitation of a natural resource is used opportunistically for short-term purposes in order to finance welfare as well as private and public consumption. The economy becomes increasingly driven by consumption, and the generous use of easy money diminishes the immediate need for structural economic change, weakens industrial competitiveness, but benefits political stability. The long-term risk is the 'Spanish disease', where the use of a financial endowment for consumption leaves the country without an up-to-date income basis and without money.

Resource risk and resource opportunity

The international debate tends to focus on the negative aspects of unusual resource wealth, using the term 'resource curse' (Humphreys *et al.*, 2012). Among the various resources, oil has been named the curse *par excellence*. The implicit proposition put forward by many discussants is that a generous natural resource endowment, in particular of oil, will leave the countries in question worse off in the end than if they had been without the resource endowment. This reasoning applies particularly to developing countries which, due to discoveries or abrupt changes in international markets and prices, suddenly find themselves with huge revenues. The criticism is briefly that these countries got too much money too quickly and squandered the funds.

The unexpected flow of money is essentially rentier income, extraordinary profit due to a random combination of shifting circumstances. The resource rent appears as a gift from nature. The issue is how this gift is handled, by whom and for what purpose. The inflow of rentier money distorts established economic circuits and provides the rulers with money without taxing the population or contracting loans. The outcome depends on conscious policy choice and on the balance of political power in the countries concerned. Experiences vary, not only between developing and industrial countries, but also among developing countries as well as among industrial countries. Among the former, Venezuela is different to Bolivia, Iran or Saudi Arabia; among the latter, Norway is different to Canada or the United Kingdom. There is no cast-iron law that resource endowment is ultimately a curse, although it can be a sedative. Russia is a case in point.

The risk is that the flows of rentier income will initially cause surpluses, drive up the exchange rate, lead to cost inflation, boost imports, depress the rest of the economy, and in the longer run lead to unemployment, a depreciating currency that causes price inflation, budget and current account deficits, social unrest and political instability (Humphrey *et al.*, 2012).

Insofar as these efforts have been financed by oil and gas revenues, it would be misleading to assert that resource extraction has not had lasting, beneficial effects. This does not imply that benefits could not have been higher, but it is doubtful whether other economic activities, without an element of economic rent, would have provided corresponding resources and certainly not as easily. Without oil and gas revenues, a corresponding investment effort would have required higher savings rates and lower consumption, possibly also higher borrowing. The argument that without oil and gas revenues the countries concerned would have been better off is hard to substantiate. When facing unexpectedly high income, the challenge is to implement policies that favour investment over consumption.

References

Brailovsky, V. (1981). Industrialisation and Oil in Mexico. Barker, T. and Brailovsky, V. (eds.). *Oil or Industry?* London: Academic Press, 71–134.

Bret-Rouzaut, N. and Thom, M. (2005). Technology Strategy in the Upstream Petroleum Supply Chain. *Les cahiers de l'économie*, 57. Rueil-Malmaison: Institut Français de Pétrole.

Bjerkholt, O., Lorentsen, L. and Strøm, S. (1981). Using Oil and Gas Revenues: The Norwegian Case. In T. Barke and V. Brailovsky (Eds.), *Oil or Industry?* London: Academic Press (1981), pp. 171–184.

Chevalier, J-M. (2005). *Les grandes batailles de l'énergie*, Paris: Gallimard.

Cleary, P. (2016). *Trillion Dollar Baby*, London: Biteback Publishers.

Ebrahim-Zadeh, C. (2003). Back to Basics. *Finance and Development*, 40(1).

Eika, T., Prestmo, J., and Tvete, E. (2010). *Etterspørselen fra petroleumsvirksomheten – Betydningen for produksjon og sysselsetting i Norge*. Statistisk Sentralbyrå, Økonomiske Analyser 2010 (3), Oslo.

Ellman, M. 1981. Natural Gas, Restructuring and Re-industrailisation: The Dutch Experience of Industrial Policy, in Barker, T. and Brailovsky, V. (eds), *Oil or Industry?* London: Academic Press.

Gylfason, T. (2001). *Neither Dutch Nor Disease, Lessons from the Dutch Disease: Causes, Treatment, and Cures*. Department of Economics University of Iceland, Institute Of Economic Studies Working Paper Series W0. Reykjavik.

Herb, M. (2014). *The Wages of Oil*, London: Cornell University Press.

Humphreys, M., Sachs, J.D. and Stiglitz, J.E. (eds) (2012), *Escaping the Resource Curse*, New York: Columbia University Press.

Karl, T.L. (1997). *The Paradox of Plenty: Oil Booms and Petro-States*, Berkeley: University of California Press.

Lam, R. and Wantchekon, L. (2002). *Political Dutch Disease*, Chicago and New York: Northwestern University and New York University.

Mazzucato. M. (2015). *The Entrepreneurial State*, New York: Public Affairs.

Mehlum, H., Moene, K. and Torvik, R. (2005). *Institutions and the resource curse*, Oslo: Institute of Economics, University of Oslo.

Noreng, Ø. (1981). Petroleum Revenues and Industrial Income. In Barker, T. and Brailovsky, V. (eds), *Oil or Industry?* London: Academic Press (1981).

Noreng, Ø. (1997). *Oil and Islam*, London: Wiley.

Olson, M. (1971). *The Logic of Collective Action*, Cambridge, Mass.: Harvard University Press.

Sachs, J.D. and Warner, A.M. (2001). The Curse of Natural Resources. *European Economic Review*, 45, 827–38.

Tordo, S., Warner, M., Manzano, O.E. and Anouti, Y. (2013). *Local Content Policies in the Oil and Gas Sector*, Washington, D.C.: The World Bank.

Treat, J.E. (ed.) (1994). *Creating the High Performance International Petroleum Company*, Tulsa, Oklahoma: PennWell.

Williamson, J.G. (2003). *Was It Stolper-Samuelson, Infant Industry or Something Else?* World Trade Tariffs 1789–1938. NBER Working Paper No. 9656.

Additional sources

Integrated Oil and Gas. (2001). Sector Report, 02.2001, London; HSBC.
Oil and Gas Exploration and Production: Reserves, costs, contracts. (2004). Institut Français du Pétrole Publications, Paris: Éditions Technip, 156.
Oil and Gas Technology Development. (2007). Working Document of the NPC Global Oil & Gas Study, Washington, D.C.: American Petroleum Council.

15 Collaborative innovation in the Norwegian oil and gas industry

Surprise or sign of a new economy-wide paradigm?

Charles Sabel and Gary Herrigel

Introduction

This book tells the story of the improbable rise of internationally competitive, Norwegian suppliers of sophisticated subsea equipment, advanced seismology and other capital goods to the offshore oil industry. By venerable consensus in development economics, the production of commodities based on natural resources, whether in agriculture or mineral extraction, is the last place to look for accumulation of the general-purpose capacities that underpin economic growth. When, exceptionally, such capacities do arise in commodity production, multinational firms produce them, or swiftly wrest control from upstart innovators. Up to now, Norwegian suppliers have succeeded against the odds both because of peculiarities in the oil industry's technological trajectory, and the infant-industry protection of domestic producers. But continuing reliance on an exhaustible natural resource like oil makes this unlikely success fragile and fleeting, and prudence suggests that the supply industry diversifies into more robust pursuits before good fortune turns bad.

The story of the oil-field supply firms' success makes an important contribution to another, more general narrative: the emergence at the frontier of production in all sectors of the economy – in the production of resource-based commodities as well as industry and services – of a model of collaborative innovation. This model takes as its starting point a world – ours – in which the speed and uncertainty of technological and market change make it impossible for even the most capable actors to master by themselves the skills needed to remain at the forefront of development. Given this uncertainty even the leading firms come to depend upon shifting constellations of suppliers, large and small, to deploy and develop technologies changing so rapidly that their strengths and weaknesses can only be determined in use.

In this new model of production, the master skill is the capacity to collaborate in adjusting general technologies to particular contexts, and then learn from those adjustments how to advance general understanding. Because this model of production is inherently collaborative, its development goes hand in hand with the creation of new forms of contract that facilitate joint innovation among rapidly changing groups of partners. Because this model of production is taking hold in all sectors of the economy, it implies that the possibilities for acquiring the general purpose capacities that are the building blocks of growth are broader than development economics has assumed.

The upshot is that firms and (parts of) national economies that excel in these disciplines of collaborative innovation thrive, regardless of whether formal ownership of productive assets is national or foreign, and regardless of whether those assets are deployed in natural-resource-based activities or not. From the perspective of this emergent model the success of the Norwegian supply industry is not an accidental exception to the exceptionally unpromising conditions of skill acquisition in a resource-based industry in a small country. It is rather a paradigmatic illustration of how the collaborative capacity to use ever deeper knowledge of the local circumstances of production can become a springboard to lasting global success.

The argument proceeds in two steps. First we recall the incremental, but cumulatively transformative development of the technology of off-shore oil extraction in which Norwegian capital goods suppliers have played an important part. Both the summary statistics for the global industry and the detailed studies of Norwegian developments presented in this book show that the epicentre of innovation is no longer the international oil companies but rather the suppliers of oilfield services and equipment.

These changes are of a piece with developments in industries as different as automobiles and pharmaceuticals. In these industries too, the locus of innovation has shifted. For much of the last century, leading firms in these and many other industries dominated research and development in their domains, and produced key components and other inputs themselves or under contract with suppliers working to detailed specifications they provided. In this sense, they were vertically integrated, whether they formally owned their suppliers or not. As in the oil case, these industries have, in their own way, also undergone technological transformations so broad in scope and unpredictable in direction that no firm can expect to be a pioneer in all. Firms that owned their suppliers have divested them out of fear that the internal producers could easily become obsolete or superfluous; and lead firms expect to collaborate with new generations of independent suppliers in developing innovative technologies.

To understand how collaboration is organized under these conditions we look, in a second step, at the contractual relations governing the co-development of innovative products in these same industries. Here we encounter a paradox. Lawyers understand contract as an exchange of precise promises: I give, so that you will give in return – *do ut des*. But when no one can say for sure what is feasible – when innovation is necessary – a detailed and binding division of labour would be a self-deluded and self-defeating speculation. The same uncertainty or inability to anticipate future states of the world that makes collaboration between independent parties indispensable to innovation makes it impossible for conventional contracting techniques to manage the effort.

Faced with this dilemma firms have adopted novel forms of interacting. Instead of attempting to specify the outcome – what each party must actually contribute to realize the common purpose – they agree to set terms for collaboration. The formal contracts create governance processes – for instance regular meetings to review progress towards milestones, and procedures for resolving conflicts over how and whether to proceed – that allow each party to assess the capacities and good faith of the other while refining ideas of what the eventual product should be – or whether it is feasible at all.

The Norwegian oil industry has adopted 'front end engineering and design' (FEED) contracts of this general type to govern collaboration with large general contractors in designing the system of production in a particular field. This is a crucial stage of development when the innovative integration, in view of basin-specific conditions, of established and novel technologies can determine the profitability of a venture – and trigger further cascades of technical innovation as ambitious designs are realized. We discuss these relatively recent and little studied contracts against the backdrop of closely related innovations in contracting in the pharmaceutical and automobile industries.

Norway, a small country on the European periphery and long dependent on the export of natural resources, has a fertile tradition of development economics (Nurkse, 2009). Like the larger body of thought with which it is intertwined, that tradition sees manufacturing industry as the high road to the acquisition of general purpose capacities, and the exploitation of natural resources as more of a curse than a blessing. The similarities between the emergent patterns of innovation in oil and gas, automobiles and pharmaceuticals is at odds with this tradition of thought. By way of conclusion we doubt the utility, today, of distinguishing natural resources as a distinct sector of the economy.

The trend to vertical disintegration and collaboration

Developments in the oil and gas industry are not singular. They are part of a very general pattern. In one area of the economy after another – within natural-resource-based sectors and within manufacturing industry itself – the very direction of technological change has become unforeseeable. Knowledge of what works today is an unreliable guide to what will be required tomorrow. The common response to this uncertainty is the opening or dismantling of closed R&D and production systems and significantly more collaboration with outsiders in both.

To illustrate the broad scope of this claim, we examine the similarities between developments in oil and gas and two non-natural-resource-based industries in the United States, automobiles and pharmaceuticals. These industries differ not only from oil and gas; they are also different from each other. The US auto industry pioneered mass production and vertical integration to become an emblem of traditional manufacturing, while pharmaceuticals have become, in the wake of the biotechnology revolution, an emblem of the new, knowledge-based economy. We find that each industry has its own trajectory, but those trajectories are all marked by a degree of uncertainty that leads to vertical disintegration and collaborative innovation. The common response suggests a change in the general conditions of competition, not the idiosyncrasies of any particular pursuit.

Collaborative innovation in Norwegian oil and gas

The extraction of oil and gas from deep reservoirs offshore under harsh conditions has been transformed in recent decades by new instrumentation and by innovations in subsea or ocean floor technology, as described in Chapters 2 and 4 of this book. One of the most important advances in metrology is 4-D or time lapse seismology, which permits observation of a reservoir as it is being drained, revealing untapped reserves. The advances in subsea technology make it possible to separate solids and seawater from oil and gas at or near the wellhead; re-inject seawater into the formation to compensate for the decrease in pressure caused by the well flow; compress gas; and pump only the gas and oil – purified of the by-products of extraction, and often together, in a single, multiphase pipeline – to floating storage facilities or to onshore terminals. Used together with advanced metrology such subsea equipment, installed by robot, monitored from shore, and operating in the high-pressure but stable conditions beneath waves and weather dramatically lowers the cost of extracting oil deep offshore, substantially expanding the range of recoverable resources while reducing the risks of harm to human operators and the environment.

Geopolitical events reinforced the destabilizing effects of these developments on oil industry organization. Price shocks in the 1970s led the IOCs to reduce expenditures on research and development, and outsource many technically demanding tasks to oil service providers. The rise of national oil companies displaced the IOCs from reserves they had long exploited, and forced them to explore alternatives in less hospitable settings, whose accessibility often depended on technical innovation. Technological and geopolitical turbulence has made the IOCs more dependent on innovative suppliers and created opportunities for Norwegian firms – as pioneers in both 4-D seismology and subsea processing – to establish themselves first in the North Sea and then internationally.

The growing contribution of suppliers to technological development in the industry is significant enough to show up in aggregate statistics, despite many problems of measurement (Chapter 3). A recent survey of senior managers in the oil and gas industry globally found that 63 per cent of the innovations actually put to use originated in service companies (Perrons, 2014). Before the IOCs began to rely on outside suppliers to (co)develop technology, the 11 leading firms in the industry accounted for more than 80 per cent of total investment in research and development (Perrons, 2014). By the mid-1990s observers found that technology had 'become so sophisticated, broad, and expensive that even the largest companies can't afford to do it all themselves' (Leonard-Barton, 1995, p. 135). A dramatic increase in the market capitalization of service and technology providers is a further indication of their increasing centrality to the industry: Schlumberger, the biggest of them, has a market value greater than all but the very largest IOCs (Perrons, 2014).

The study of Norwegian suppliers (Chapter 3) reflects this trend. Oil-related suppliers outperform Norwegian firms generally and Norwegian firms in comparable sectors. In particular, the oil field suppliers scored high on measures of participation in research and development projects and less formalized collaboration with customers and other suppliers.

Social network analysis of all publicly funded, oil-related, Norwegian research projects shows these relations in higher resolution. In projects regarding the implementation of new technology on the Norwegian continental shelf (NCS), the supplier company is typically the project leader, and other actors cluster around it. Collaboration among suppliers is increasing. In short, suppliers play a lead role in innovation on the NCS (and by extension in global markets) and their reliance on each other is increasing.

Interviews with supplier managers from the Rogaland (Stavanger) subsea technology and service cluster and their large oil operator customer counterparts confirm the centrality of suppliers to innovation (Chapter 4). Virtually all the supply-firm managers agree that in-house knowledge accounted for at least 65 per cent of innovative solutions. The oil operating company managers, interviewed less systematically, concurred, estimating that their companies contributed only between 15 and 35 per cent of the know-how that goes into collaborative innovations.

The precise balance of the contributions matters less here than the nearly unanimous agreement that innovation is no longer the exclusive domain of the oil producers. Innovation today is, instead, vertically disintegrated and collaborative, with suppliers often taking the lead in joint efforts. A brief look at changes in the automobile and pharmaceutical industries shows that this is a general outcome, even if the paths to it vary by industry.

Vertical disintegration in automobiles

The US automobile industry is a particularly useful reference point because it was for much of the twentieth century the canonical example of the natural connection between mass production and vertical integration. The parts of a car were specialized to one another: each was good for a particular make or model. Manufacturers, fearing supply disruptions – particularly costly because production was capital intensive – made the bulk of almost all the parts they needed. They only resorted to outside suppliers for two reasons. One was to periodically check that the methods and technologies used internally were at least as efficient as those available on the outside market, and to take corrective action if not. The other was to hedge against unpredictable, short-term fluctuations in demand: outside supply could be increased or decreased as needed to ensure that internal capacity was utilized at a nearly constant, efficient rate.

The kind of geopolitical shocks that broke open the oil industry in the 1970s deeply unsettled US automobile manufacturers as well. Yet, despite fitful efforts to emulate Japanese competitors who were then demonstrating the potential of collaborative methods, the initial and for decades main US auto company response was simply outsourcing: divesting internal units and purchasing parts and components from independent suppliers, still working to detailed specifications, and often located in low-wage countries. Disintegration was dramatic and relentless. In the early 1980s, as outsourcing began, General Motors, the largest of the US companies, produced roughly 70 per cent of its parts and components internally, and contracted for the remainder; by the early 2000s the ratio was reversed. This was, however, vertical disintegration in the narrow sense of increased reliance on outside markets but not in the larger sense of the term as including a shift to collaborative innovation: the automakers continued to dominate design and development in the global supply chains they created.

This hierarchical disintegration began to give way to increasingly collaborative relations after the turn of the century; and the process picked up speed as the global economy recovered from the financial crisis. As with the oil producers in roughly the same period, the automakers' inability to keep abreast of the proliferation of transformative innovations ultimately undermined hierarchical control over technology in the supply chain. The growing importance of microprocessors in engine control and safety equipment such as airbags and anti-lock brakes, for example, made them increasingly dependent on novel technologies beyond their ken, and therefore on collaboration with outside suppliers who had mastered them. More recently collaboration with battery makers and the producers of other key components for electric cars crystalized the changing relation between the automakers and their suppliers. Whereas the supplier's traditional aim was to win the preference of one major customer, the situation is now reversed. It is the buyers who must compete for the preference of capable suppliers. As one manager puts it, becoming the supplier's 'customer of choice is increasingly an imperative' to assure the partner's best efforts in collaborative innovation (Trebilcock, 2017, 21).

... and in pharmaceuticals

Where the shift to collaborative innovation came incrementally in oil and autos, in pharmaceuticals it came almost explosively, as the result of a technology shock. Large pharmaceutical firms built in-house research laboratories to improve the commercial production of penicillin before and during the Second World War. After the war these facilities were used to screen natural and chemically derived compounds for potential therapeutic activity. Progress in physiology, enzymology, cell biology led to deeper understanding of the biochemistry and molecular biology of many diseases, and insight into the mechanism of action by which some drugs achieved their effects: random screening gave way to 'guided search' and 'rational drug design' as the precepts of investigation, but research continued to be highly centralized in large pharmaceutical companies (Malerba and Orsenigo, 1992).

It was the following wave of innovation, in genomics, genetic engineering as well as molecular biology, that, from the late 1970s on, ended the dominance of the internal research units and opened the way to collaborative investigation. The new science made it possible to grasp with much greater precision – though still incompletely – the chains of biochemical reaction by which particular cancerous tumours propagated, or sheltered themselves from immunological defences – and to devise counterstrategies. By the 1990s, universities, public research organizations, venture capitalists and successive cohorts of new biotech firms were collaborating and competing with multinational pharmaceutical corporations to develop and produce the new technologies.

Powell and his collaborators have used network analysis to trace the rapid evolution of the ties (including research and development collaborations, financing, and commercialization and licensing arrangements) among these various actors from 1988 to 1999 (Powell *et al.*, 2005). In the early period there is a relatively clear division of labour. Small biotech firms, often start-ups, do innovative research; multinationals and first-generation biotech firms help finance the research and commercialize it by organizing the expensive tests on successively larger groups of human subjects needed to demonstrate the safety and efficacy of new drugs to regulators. There is a continual flow of 'new entrants as progress is made along a broad scientific frontier in which no single organization can develop a full range of scientific, managerial, and organizational skills' (Powell *et al.*, 2005, p. 1188). As time goes on the actors with the most connections, at the centre of the network, are those with the most diverse set of collaborations, suggesting 'an open elite, accessible to novelty as the field expands' (Powell *et al.*, 2005, p. 1189). The 'large' players, in other words, do not become central because of the number of their connections, but because they are good at playing, collaboratively, with others.

Leading figures in the oil and gas industry have noticed these trends toward vertical disintegration and increasingly close collaboration with suppliers and urged their peers to learn from them. A former chairman and CEO of Schlumberger, for instance, remarked recently that firms in the automobile and mobile phones industries had been quicker than firms in his own to 'recognize their subcontractors are an essential part of the product or product development, and involving them in a closer partnership relationship is an essential competitive advantage' (Finding petroleum, 2012). As we will see in the next section this is just the direction in which contracting with suppliers in the oil and gas industry is going.

Contracting for innovation

The sociology of work and industry is congenitally suspicious of contracts or formal agreements as descriptions of how work is done. A finding so common that it became an assumption is that even the most apparently routine tasks can only be accomplished by informal adjustments unforeseen in the instructions for the job (Fox, 1974). Because this is so, following rules to the letter is, as trade unions discovered, an ingenious strategy for stopping work while complying with formal obligations. Companion research in law and society found similarly that contracts between small and medium-sized firms are often a mere formality (Macaulay, 1963). When disagreements arise the parties react on the basis of shared norms of reciprocity – trust – inflected by understandings of their own history of past dealings. The contract stays, forgotten, in the drawer.

This research implies that as uncertainty increases, and with it the need for collaboration under unforeseeable circumstances, trust will become more important as the foundation of cooperation, and contracts even less. Thus successful collaboration between Japanese automakers and their suppliers – instead of vertical integration – was often attributed to a national culture of reciprocity that mitigated the risks of opportunism (Holstrom and Roberts, 1998). Conversely it has been argued that the continuing failure of US automakers to adopt the plainly superior Japanese methods is rooted in the persistence in the US industry of an unwritten, relational contract – a de facto social pact – that grew from old practices and obstructs the introduction of new ones (Gibbons and Henderson, 2012; Helper and Henderson, 2014).

But there is paradox here too. As the trajectories of technological development become less predictable, and results in one domain prove applicable far afield, collaboration will often be between near strangers, who presumably do not share social norms. 'Open elite' at the centre of the biotech networks – the most detailed picture of collaboration under uncertainty that we have – succeeds because it remains 'accessible to novelty as the field expands' – committed to seeking out and working with promising strangers. As the level of uncertainty rises, moreover, the nature of tasks change and past performance is a poor predictor of the capacity to meet new obligations. Established partners may intend to meet their responsibilities, but be unable to do so. The very uncertainty that makes collaboration more important thwarts not only traditional contracting but also pre-contractual reliance on social norms as a coordinating mechanism as well.

The novel contracts and contract governance mechanisms (we explain the difference in a moment) emerging in the oil, biotech and automobile industries are an apparently successful legal adaption to the new constellation of requirements. Compared to unspoken relational contracts they are remarkably formal. They require the parties to set general goals; define milestones on the way to them; review interim results regularly to identify problems; redirect efforts to revised goals – and perhaps abandon the project. In egregious cases of 'red-faced cheating' – secretly channelling the ongoing results of collaboration with an external partner to a parallel internal research unit – courts will sanction breaches of the agreement.

But compared to traditional contracts they seem like informal arrangements, not contracts at all. The parties' only obligation is to collaborate without engaging in red-faced cheating. Neither party must produce any determinate result, either in the course of the collaboration or, *a fortiori*, at the end of it. Even collaboration that results in a potential product creates no further obligations.

The buyer will have an actual or de facto option to purchase the result, but can choose not to exercise it (in which case the seller will have an actual or de facto option to make use of the product).

How can such non-binding agreements further collaboration under the demanding conditions of uncertainty? The answer is by using the formal requirements of explicit collaboration to generate the informal norms that make mutual dependence under changing circumstances workable and tolerable. The information exchanged under the formal contractual provisions allows the parties to evaluate one another's capacities and good faith – to observe if the capable stranger can become a reliable partner and the trusted partner is capable of new tasks – while simultaneously evaluating the prospects of the particular project and joint efforts generally. As collaboration progresses each party relies increasingly on the capacities of the other, deterring opportunistic defection even in the absence of an explicit commitment to purchase anything in advance. In this way the formal contractual obligations – regular review and deliberate consideration of the interim results – create the conditions in which informal norms and self-interested calculations bind the parties to continue promising collaboration in good faith. Because they have the explicit aim of encouraging collaboration whose outcome cannot not be defined in advance, we will call contracts of this general kind contracting for innovation (Gilson *et al.*, 2009).

Pharmaceuticals

As the shock of technological innovation came earlier and more abruptly in pharmaceuticals than in oil or automobiles, it is not surprising that contracting for innovation is most developed there. In a typical agreement, a large pharmaceutical company with, say, expertise in a particular pathology's metabolic pathways, collaborates in a search for therapeutic compounds with a small biotech company that has developed tools for identifying molecular classes likely to correct the metabolic defect without producing toxic side effects. If a promising compound is found, the large pharma company takes charge of clinical testing and other regulatory obligations.

Typically the agreement provides for a research period of roughly a year, which may be extended by agreement of the parties, to fully explore the possibilities of a solution. During this period the large firm will usually pay the smaller partner's research costs, and make additional payments for achieving agreed milestones. The agreement allows for searching, and joint evaluation of the research results, yet avoids deadlock and protects both parties against opportunist use of the findings – without creating a binding obligation to produce any result at all.

A key institution created by these agreements is a joint research committee, composed of three or so managers or researchers from each of the partners, all directly involved in the project. This committee recommends continuation of the project, or not, during the research period, or extension of the work after the anticipated endpoint. Recommendations to continue require unanimity, so committee members with doubts about the project can easily request further information. In the case of deadlock, decisions are escalated to a high manager in each company. Knowing little of the day-to-day operation of the project, these managers make decisions on the basis of the record produced by the committee's deliberations. Holdouts who object without clearly demonstrable reasons run the risk of being disavowed by their superiors – a prospect that deters self-interested obstinacy. If the big pharma company disregards the committee's recommendation for commercial reasons the agreement bars it from pursuing further similar research for an extended period. The pharma company has an option to purchase the rights to further development (paying a licence fee to the biotech firm in case of ultimate success). If the pharma company does not proceed, the biotech firm has a matching option to acquire the development rights at a lesser cost.

The US auto industry

The shift towards systematic collaboration came in the US auto industry not through revision of the terms of the contracts – which have remained lopsidedly in favour of the buyers for decades – but rather through the introduction of a new contract governance regime: rules for reciprocal performance review and procedures for responding to the problems and opportunities detected that together transform the effective operation of the underlying agreement.

Until very recently US automakers, we saw, shifted all risks to suppliers, while buffering themselves against disruptions of supply. The 'blanket order' or master contract is used to assign the risk. Under a blanket order the supplier agrees to provide a specific part, at an agreed price. But the blanket order does not obligate the buyer to actually purchase anything. That obligation is created only when the automaker issues, often month by month, a purchase order for a certain number of parts, at the agreed price, for delivery during the following period. The supplier thus absorbs all the costs of fluctuations in demand (Macaulay, 1974). These practices survived the oil price shocks of the 1970s and the vertical disintegration that followed (Ben-Shahar and White, 2005).

But under the surface there was change. Japanese transplants' success working with US suppliers demonstrated both the benefits of collaboration and domestic firms' aptitude for it. Moreover, the differences in buyers' capacity for and dedication to collaborative problem-solving and improvement – and their willingness to share the resulting gains – were becoming public knowledge. By the turn of the century suppliers plainly preferred some buyers to others. Beginning in 2001, an annual survey of suppliers' experience with their customers was used to produce the Supplier Working Relation Index (SWRI), which publicly ranks buyer performance.

As the pace of technological innovation accelerated after the financial crisis, and with it the need for collaborative innovation with capable suppliers, GM – at the bottom of the SWRI rankings – deliberately decided to establish itself as a reliable partner or preferred customer. Adding contractual language obligating the parties to make their best efforts in collaboration was impractical. Except in egregious cases, GM would not be able to verify to a court that a supplier had stinted in its efforts or withheld (emergent and tentative) understanding of technological prospects.

Instead GM introduced an information exchange regime, resembling the one centred on joint project committees in biotech, that makes the parties' capacities and intentions easily observable, even if not legally verifiable. In the new governance regime, GM gives suppliers monthly reports comparing target and actual performance on current operation measures such as quality and material cost, and annual reports on measures of collaborative performance such as responsiveness and engineering. Suppliers review GM in the same way. This regular, mutual review promptly alerts each party to the other's problems and allows both to make informed evaluations of the responses.

These measures produced a turnabout in GMs standing as a buyer, moving it from tied to last in the SWRI rankings to within striking distance of the leaders, Toyota and Honda. GM improved moreover in five of the six procurement areas included in the survey, providing further evidence of corporate-wide change (Henke, 2017). As the ratings rose, suppliers reduced their selling prices to GM, indicating that the new relation was yielding mutually beneficial efficiency gains (Trebilcock, 2017).

These results suggest that the information exchange regime, based on mutual monitoring, is the engine of contracting under uncertainty. The traditional system of non-binding, long-term blanket orders and binding but short-term purchase orders continues as it has historically. Unwritten, relational contracts cannot, by their nature, change in the short term. The only significant change in the relation between GM and its suppliers – and the most plausible explanation of improved collaboration – is the introduction of a highly formalized regime for continual reciprocal review, designed to induce, and braid with, informal mechanisms for resolving eventual disagreements – contracting for innovation.

Contracting for innovation in the Norwegian oil industry

As noted at the outset, changes in the methods and organization of the oil industry generally, and in Norway in particular, have been incremental. There was no single technological shock equivalent to the eruption of biotech in pharmaceuticals or the advent much more recently of electric cars in the auto industry. But cumulatively incremental change has produced wave after wave of reorganization in the oil industry, with corresponding changes in the forms of contracting, though earlier forms often coexist with later ones.

For the sake of exposition we divide oil operator – supply/service company developments from the 1980s to the present into three periods: an initial phase of vertical integration in our expanded sense of the term, where the operators direct the suppliers; a turn, beginning in the mid-1990s, to turnkey contracts, where a lead supplier or general contractor takes responsibility for coordinating the work of the others; and most recently, as recognition grows that the initial or front-end designs increasingly determine the economic viability of demanding projects, formalization of this stage of the design process in FEED contracts as a distinct, standalone and open-ended collaboration between the operator and a qualified partner. It is in this third and most recent phase of development, as the parties deliberately explore innovative solutions to highly specific problems without committing themselves to executing the project together, or to proceeding at all, that contracting for innovation comes to the oil industry.

In the first, vertical integration, phase, in the 1980s, the operator's engineering staff produced a technical description of necessary equipment. An outside engineering firm drew up detailed drawings based on this description; the operator then put the drawings out to bid, selected suppliers and scrupulously monitored the execution of the plans, modifying them when necessary. To minimize the burdens of supervision and coordination major components were built sequentially, in the order of installation (Nelson and Braadland, 2014).

Nominally contracts fixed the relations between operators and suppliers. But given the inherent uncertainties of the situation – the limits to the *ex ante* specification of the final product; great imprecision in the estimation of design and production costs; and the difficulties of tracking and measuring performance – these were, as Stinchombe observed in a well-known study at the time, legal hybrids or 'chimeras' – contracts as hierarchical documents (Stinchcombe, 1985). Though the agreements spoke in the language of prices and incentives familiar from contract, they included provisions for adjusting the costs, prices and quantities agreed as circumstances changed, while imposing the buyer's standard operating procedures on the supplier and providing for internal dispute resolution mechanisms in case of disagreements. The result, Stinchcombe found, is a relation between supplier and buyer that looks 'very much like a hierarchy' – an extension of the buyer's internal order to the outside party – and very little like the arm's length dealings typical of contracting in a competitive market (Stinchcombe, 1985, p. 126).

The 'supervisory' role of the operator was most burdensome in the requirement that its engineering staff approve the work of the supplier's engineers. In the most extreme cases nothing escaped review. 'Sketches, drafts, preliminary specifications and drawings, final specifications and drawings, tender documents, technical evaluations of the bidders' replies, technical changes proposed after contract start, and costs for all of these', Stinchcombe found, 'all have to be approved by the client' (Stinchcombe, 1985, p. 157). These procedures, costly in themselves, also obscured responsibility for any decision. So complete was the system of approvals, Stinchcombe observed, that 'every step on the way to an engineering contractor's default has been approved and/or caused by a client directive', making 'the contractor's responsibility for defaults very difficult to prove in court' (Stinchcombe, 1985, p. 154). The very provisions of the contract that made them useful as instruments for collaboration under uncertainty rendered them useless as formal contracts.

This vertical integration model was workable, barely, so long as the price of oil was high and developments on the NCS were in some measure subsidized by infant industry protections. By the early 1990s, as competitive conditions tightened, the government convened major actors in the oil and gas industry, including the main trade association representatives of leading firms and regulatory authorities, to investigate structural problems in the industry and suggest solutions. In addition to proposing the development of new technical standards for all firms operating on the NCS, this forum suggested a radical reform of supplier relations and the associated forms of contracting, which rapidly took hold.

The new turn-key model, and the engineering, procurement and construction (EPC) contracts in which it was embodied, was in many ways the reverse of the previous one. Instead of selecting a series of suppliers, one after another, and carefully monitoring each as it executed detailed design drawings, the operator in the EPC model entered a single, comprehensive agreement with a capable systems integrator. The systems integrator then assumed responsibility for many of the tasks that had fallen to the operator in the vertical integration model: developing detailed plans from the functional specification of equipment requirements, selecting suppliers and supervising their work. Execution was now simultaneous, not sequential, so the systems integrator had responsibility for coordinating relations among suppliers with complementary tasks as well.

The EPC contracts seem to have contained the escalating supervisory costs of the vertical integration model, but the turn-key approach had limits of its own. Few systems integrators were capable, diversified and deep-pocketed enough to assume the risk of managing extremely large, complex and uncertain projects, even with the addition of contractual provisions limiting their liability to some share of the total project costs. The limited competition for EPC contracts reduced the pressure on those firms that did enter bids to keep prices low, and especially to contain demands for large risk premiums as a protection against costly, unforeseen developments (Nelson and Braadland, 2014).

The costs of uncertainty have proven to be an ongoing source of consternation in EPC contracts. Precisely because the systems integrator is in control of the project from early on, detection of what the operator sees as problems may be delayed; and delay can substantially increase the costs of modifications and trigger recriminations about liability. The standard form EPC contracts therefore elaborately specify the system integrator's obligations to notify the client of difficulties in order to escape or limit liability – an indication that the 'supervisory' problems of the hierarchical model had been transformed, not definitively resolved (Beidvik, 2011).

In practice the operation of the EPC-contracts seems to depend, as the older sociology of industry would lead us to expect, on the prior relations of the participating firms and the ad hoc governance arrangements they create to manage particular projects. When the participants are familiar with each other and on easy terms – when trust is high – they may agree to form an 'alliance' or some other arrangement for pooling project experience, identifying problems and devising joint solutions. When trust is low they may create multiple and competing governance mechanisms, which each participant or faction uses to advance its interest against the others (Olsen *et al.*, 2005).

But these idiosyncrasies aside, as novel technologies are deployed in harsher and harsher environments on the NCS, EPCs appear to be reaching their limit as the master coordination instruments in large oil projects. Two related problems stand out. The opening 'pre-project' or FEED stage of development, where innovative, basin-specific solutions are thoroughly explored, is becoming increasingly important to a project's overall success. A study of significant cost overruns and delays in large off-shore projects traced their cause back to deficiencies in this stage: faulty or incomplete exploration of possible solutions evidently cannot be compensated by improvisation later; on the contrary, initial shortcomings cascade, increasing costs and delaying completion (Oljedirektoratet, 2013). But the skills needed for this new wider-ranging FEED exploration are not obviously those of the traditional systems integrator. Indeed, as the range of technologies implicated increases and thorough evaluation of them becomes more crucial to a project's success, the ability of traditional turn-key experts to lead the decisive pre-project work comes into question. Their habitual approaches to problems and long-standing connections to suppliers, often chosen more for their reliability than for their innovative capacity, leave them underequipped for the more demanding and collaborative exploratory processes.

The third and most recent development in contracting in the Norwegian oil and gas industry begins to address both problems in ways that recall the contracting for innovation models in pharmaceuticals and autos. The key change is to make FEED an independent project or group of projects, rather than a stage in an integrated EPC contract. Under vertical integration the operator was chiefly responsible for design decisions; in the EPC model the systems integrator has substantial, often decisive control. In this emerging third model, operator, integrator and specialized suppliers all aim to collaborate fully in investigating possibilities. FEED projects can be executed in sequence, with each project refining the more general design produced by the preceding one. They can be nested as well, so that a FEED project examining overarching designs gives rise to FEED evaluations of various components on which possible solutions will depend. Besides allowing for a more comprehensive and thorough canvass of alternatives, treating FEED as a standalone project helps reduce the likelihood that conceptualization of a solution is subtly influenced by consideration of the capacities of the firms that will execute it. The commitment to collaborate in the search for solutions does not entail a commitment to act together on the results. Once a plan is devised, the operator is free to enter an EPC contract with the FEED partner, choose another turn-key provider or oversee some aspects of the project directly while delegating responsibility for the others by entering one or more EPC contracts of limited scope.

To all appearances FEED projects have been proliferating in recent years, and are playing an increasingly prominent role in project design. Observers agree that they involve close collaboration between operators and FEED teams, often with participation of key specialist suppliers. 'Early involvement' and 'tight collaboration' with suppliers in the framework of a FEED agreement is credited with halving the capital costs of developing the Johan Castberg field in the Barents Sea, lowering the breakeven price of recovering oil in the reservoir from $80 to $35 per barrel (Barstad, 2017). Use of FEED contracts yielded similar reductions in capital expenditure in the giant Johan Sverdrup field in the North Sea, again because of the early involvement of key actors and improved handoffs from one stage of the project to the next (Statoil, 2013).

Though these results are suggestive, we still lack detailed accounts of the institutional mechanisms by which the collaborative exploration of uncertainty is organized. These might be set out in the formal terms of the FEED contracts, on the lines of the research agreements in biotech, or contained in the governance routines that determine how the contracts are applied in practice, as in the recent developments in the US auto industry. If, as we are strongly inclined to believe, the introduction and spread of FEED contracts in the Norwegian oil and gas industry marks a shift towards contracting for innovation, we would expect to see that collaboration depends not primarily on trust born of prior association but, instead, on the ongoing mutual review of performance and joint resolution of the problems this scrutiny reveals.

Conclusion: turning the page in development economics?

That innovation in the Norwegian oil and gas industry closely resembles the pattern in automobiles and pharmaceuticals is further evidence, if more is needed, that it is time to turn the page in development economics and rethink the role of natural resources in growth. The idea that distinguishes development economics as a discipline is that manufacturing industry, alone among economic activities, allows, indeed induces, the accumulation of general skills through learning by doing. Once firms in a developing economy enter the market for particular industrial goods, their productivity is assumed to inexorably increase, resulting in 'unconditional convergence' with the frontier of performance. In natural-resource based sectors and services, in contrast, innovation is infrequent and usually the product of breakthroughs by powerful foreign actors, not local learning. It is a short step to the conclusion that natural resources, in the form of mineral deposits or arable land, are more often a curse than a blessing, stunting learning, subjecting a developing economy to the dictates of international capital markets and – in the case of oil bonanzas and the like – attracting waves of inward investment that discourage domestic manufacturing by raising wages and overvaluing the exchange rate.

The idea of unconditional convergence is literally history: it was true of earlier cohorts of industrializers, such as Japan and Germany, but it is not true of more recent ones, including China, a manufacturing giant (Diao *et al.*, 2017). It is well established that industry plays a smaller role in modernization of developing economies today than it did in the past (Rodrik, 2016). It is a commonplace misconception that entering many low-skill industrial activities leads nowhere. The notion of a resource curse has been refuted so many times that it will soon enter that intellectual nether world where ideas are invoked only to be rejected (Hallward-Driemeier and Nayyar, 2017).

But the study of innovation in the oil industry compels us to go further. The striking similarities between developments there – and by extension in other natural-resource based activities – and in very different industries, old and new, defy the claim that today, at least, there is any fundamental difference between them. In all, innovation is so pervasive and fraught with uncertainty that it can only be mastered by collaboration – and by collaboration institutionalized in similar ways.

These findings imply that the idea of natural resources as a separate compartment of activity is a false distinction. As the increasing rate and scope of recovery on the NCS shows, the application of knowledge to natural resources produces new knowledge – and new resources. It is as misleading to think of those resources as given once and for all by a nature beyond our reach as to think of knowledge as renewing itself in a separate world of abstractions sealed away from material entanglements. Knowledge creates resources just as experience creates knowledge (Ville and Wicken, 2012; David and Wright, 1997).

This continuing to and fro between theoretical word and practical deeds today takes the form of collaborative innovation. Contracting for innovation is one of its principal instruments. It allows trusted partners to verify that the capacities they have are the ones needed, and to renew them if not; it allows relative strangers to build trust as they explore uncertainty together. It allows partners, old and new, to bind themselves legally without entangling cooperation in formalities that defeat collaboration. Contracting for innovation is thus a kind of flexible joint or connector that facilitates the recombination of the pieces of the learning economy across all sectors.

Taken together the breakdown in the distinction between natural-resource based activities and the rest of the economy and the rise of collaborative innovation in response to increasing uncertainty make the prospects for development both more and less forbidding than development economics suggests. The conditions are more forbidding because there is no longer a manufacturing highroad to development – difficult perhaps to enter, but effortless to travel – and because trusted partners may prove unreliable while the need to rely on strangers increases. But the conditions of development are less forbidding because now there are as many paths to growth as there are areas of economic activity, even if none is an easy passage, and cooperation with strangers can be made less risky, and more likely to lead to trust than habit tells us.

Norway, by the quirks of its history, has been a pioneer in exploring the landscape of our new possibilities, and developments there, by turns surprising and predictable, help show us the way to make the most of them.

References

Beidvik, O. (2011). Leverandørens varslingsplikter etter NF 07/NTK 07. Konsekvensen av manglende varsel. MA, Faculty of Law. Oslo: University of Oslo.

Ben-Shahar, O. and White, J. (2005). Boilerplate and economic power in auto manufacturing contracts. *Mich. L. Rev.* 104: 953.

David, A.P. and Wright, G. (1997). Increasing returns and the genesis of American resource abundance. *Industrial and Corporate Change* 6(2): 203–245.

Diao, X., McMillan, M. and Rodrik, D. (2017). *The recent growth boom in developing economies: A structural change perspective.* No. w23132. Cambridge, Mass: National Bureau of Economic Research.

Fox, A. (1974). *Beyond contract: Work, power and trust relations.* London: Faber & Faber.

Gibbons, R. and Henderson, R. (2012). Relational contracts and organizational capabilities. *Organization Science* 23(5): 1350–1364.

Gilson, R.J., Sabel, C.F. and Scott, R.E. (2009). Contracting for innovation: vertical disintegration and interfirm collaboration. *Colum. L. Rev.* 109: 431.

Hallward-Driemeier, M. and Nayyar, G. (2017). *Trouble in the Making?: The Future of Manufacturing-led Development.* World Bank Publications, 2017.

Helper, S., and Henderson, R. (2014). Management Practices, Relational Contracts, and the Decline of General Motors. *Journal of Economic Perspectives*, 28(1), 49–72.

Holmstrom, B. and Roberts, J. (1998). The boundaries of the firm revisited. *Journal of Economic perspectives* 12(4): 73–94.

Leonard-Barton, D. (1995). *Wellspring of knowledge.* Boston, MA: Harvard Business School Press.

Macaulay, S. (1963). Non-contractual relations in business: A preliminary study. *American sociological review*: 55–67.

Macaulay, S. (1974). The Standardized Contracts of United States Automobile Manufacturers. *3–21 Int'l Encyclopedia Comp. L. 18.*

Malerba, F. and Orsenigo, L. (1992). Regularidades en las actividades de innovación: una investigación preliminar para cuatro países europeos. *Ekonomiaz*, 22(01), 10–29.

Nelson, T. and Braadland, M. (2014). EPC som kontraktstrategi i offshore-prosjekter [EPC as contract strategy in offshore-projects]. *Magma*, 4/2014.

Nurkse, R. (2009). *Trade and Development.* Vol. 1. London: Anthem Press.

Olsen, B.E., Haugland, S.A., Karlsen, E. and Husøy, G.J. (2005). Governance of complex procurements in the oil and gas industry. *Journal of Purchasing and Supply management*, 11(1): 1–13.

Perrons, R.K. (2014). How innovation and R&D happen in the upstream oil & gas industry: Insights from a global survey. *Journal of Petroleum Science and Engineering* 124: 301–312.

Powell, W.W., White, D.R., Koput, K.W. and Owen-Smith, J. (2005). Network dynamics and field evolution: The growth of inter-organizational collaboration in the life sciences. *American Journal of Sociology*, 110(4)): 1132–1205.

Rodrik, D. (2016). Premature deindustrialization. *Journal of Economic Growth*, 21(1): 1–33.

Stinchcombe, A. (1985). Contracts as hierarchical documents. Stinchcombe, A. and Heimer, C. (eds.), *Organization Theory and Project Management.* Bergen, Norway. Norwegian University Press.

Trebilcock, B. (2017). How They Did It: Supplier Trust at General Motors. *Supply Chain Management Review* (May/June).

Ville, S. and Wicken, O. (2012). The dynamics of resource-based economic development: evidence from Australia and Norway. *Industrial and Corporate Change* 22(5)

Additional sources

Barstad, S. (05.12.2017). Statoil bygger ut gigantfeltet til halv pris [Statoil extending giant field for half price]. *Aftenposten* (News media). Retrieved 26.02.2018 www.aftenposten.no/norge/i/P390mb/Statoil-bygger-ut-gigantfeltet-Johan-Castberg-til-halv-pris

Finding petroleum. (07.08.2012). Andrew Gould – take advantage of service industry competition. *Finding Petroleum*. Retrieved 26.02.2018 www.findingpetroleum.com/n/Andrew-Gould-take-advantage-of-service-industry-competition/07029379.aspx

Henke, J.W. (15.05.2017) 2017 N.A. Automotive OEM Study Shows General Motors Jumps to Third Place, Nissan falls to last, in Supplier Relations. *CISION PR Newswire*. Planning Perspectives Inc. Retrieved on 9/26/17. www.prnewswire.com/news-releases/2017-na-automotive-oem-study-shows-general-motors-jumps-to-third-place-nissan-falls-to-last-in-supplier-relations-300457268.html.

Oljedirektoratet [Petroleumdirectorate]. (2013). Vurdering av gjennomførte prosjekter på norsk sokkel [Assessment of excecuted projects on the Norwegian shelf]. Retrieved 28.02.18. www.npd.no/Global/Norsk/3-Publikasjoner/Rapporter/Vurdering-av-prosjekter.pdf.

Statoil. (20.12.2013). Aker Solutions tildelt rammeavtale for prosjektering på Johan Sverdrup [Aker Solutions awarded framework agreement for engeneering on Johan Sverdrup]. Retrieved 26.02.2018 www.statoil.com/no/news/archive/2013/12/20/20DecFEEDSverdrup.html.

Index

Page numbers in **bold** denote tables, those in *italics* denote figures.

For Product Safety Concerns and Information please contact our EU representative GPSR@taylorandfrancis.com Taylor & Francis Verlag GmbH, Kaufingerstraße 24, 80331 München, Germany

Batch number: 10725973

Printed by Printforce, the Netherlands